MOTION AND PAT
FOR ADDIT
MANUFACTURING

Additive Manufacturing Materials and Technologies

MOTION AND PATH PLANNING FOR ADDITIVE MANUFACTURING

ALEX C. ROSCHLI
Oak Ridge National Laboratory, Tennessee, United States
MICHAEL C. BORISH
Oak Ridge National Laboratory, Tennessee, United States
ABBY K. BARNES
Oak Ridge National Laboratory, Tennessee, United States
THOMAS A. FELDHAUSEN
Oak Ridge National Laboratory, Tennessee, United States
PETER L. WANG
Oak Ridge National Laboratory, Tennessee, United States
ERIC MACDONALD
Professor, Aerospace and Mechanical Engineering, The University of Texas at El Paso, Texas, United States

ELSEVIER

Elsevier
Radarweg 29, PO Box 211, 1000 AE Amsterdam, Netherlands
The Boulevard, Langford Lane, Kidlington, Oxford OX5 1GB, United Kingdom
50 Hampshire Street, 5th Floor, Cambridge, MA 02139, United States

ISBN: 978-0-443-15286-3

For information on all Elsevier publications
visit our website at https://www.elsevier.com/books-and-journals

Publisher: Matthew Deans
Acquisitions Editor: Dennis McGonagle
Editorial Project Manager: Mason Malloy
Production Project Manager: Anitha Sivaraj
Cover Designer: Christian Bilbow

Typeset by STRAIVE, India

Dedication

To my wife, Destinee. This book would have never happened without your constant support and encouragement. I can't thank you enough. The nights and weekends I spent writing would have been overwritten by unintelligible "writing" from Camila constantly placing her head on the keyboard.

—Alex Roschli

I'm certain if you were to have asked my family to guess about interesting life events that may happen to them that "dedication in a book for path planning for additive manufacturing" would not have even made the list. As far as they are concerned, I play on computers all day and have somehow convinced someone to pay me for it. Regardless, they still love and support me unconditionally, and without that I would not be where I am. To my wife Emily, mother Diana, and grandmother Rosemarie, thank you.

—Michael Borish

When I was about seven years old, my dad told me I should be an engineer after I insisted on assembling his new kettle grill by myself without reading the instructions. Of course, I didn't listen to him. Instead, I spent 8 years of school studying neuroscience and attempting to be a successful academic. Somehow, though, I still ended up with a job in engineering research and as an author of an engineering textbook.

To my dad, Boyd: sorry I didn't listen, but hasn't it been more fun this way?

To my mom, Elaine: your analytical mind rarely got the appreciation it deserved, but you can count this book as proof of its impact and my gratitude.

To my brother, Adam: I have always striven to be as smart as you are, and I would not be where I am today without you, whether you know it or not.

To my husband, Justin: you have given me the space, support, and fuel to grow into a person that gets to write a dedication for a textbook, and I could never thank you enough, but when have you ever asked for thank-yous anyway?

To my best friend and sister-from-another-mister, Cris: I'm pretty sure you saved me back in high school, so this one's for you, BFFFFFFFFF.

—Abby Barnes

My contribution is dedicated to my wife Cheyanne, and my children Elias James and Emilia Quinn.

—Thomas Feldhausen

My contribution is dedicated to my father William MacDonald.

—Eric MacDonald

Contents

4. Geometry data storage

Alex Roschli, Michael Borish, Abby Barnes, Eric MacDonald, Peter Wang, and Thomas Feldhausen

5. Cross-sectioning

Michael Borish, Alex Roschli, Abby Barnes, Eric MacDonald, Peter Wang, and Thomas Feldhausen

6. Closed-loop toolpath generation

Alex Roschli, Michael Borish, Abby Barnes, Thomas Feldhausen, Eric MacDonald, and Peter Wang

7. Space-filling toolpath generation

Alex Roschli, Michael Borish, Abby Barnes, Peter Wang, Thomas Feldhausen, and Eric MacDonald

Preface

I've been working as a researcher in additive manufacturing for more than a decade as part of Oak Ridge National Laboratory's (ORNL) Manufacturing Demonstration Facility (MDF). The MDF is dedicated to advanced manufacturing research with a heavy focus on additive manufacturing. In 2013, we started development of a large-format polymer pellet process that ultimately spawned the Cincinnati BAAM (Big Area Additive Manufacturing) in early 2014.

In late spring of 2014, we agreed to start a project to produce the world's first 3D printed car, live at the International Manufacturing Technology Show (IMTS) in September of 2014, using the BAAM technology. That left us just a few months to develop all of the hardware, software, and related systems to make the entire process work. As we started the project, we quickly realized that slicing software was going to be a big issue.

At the time, open-source slicing was in its infancy and provided control only for filament-based systems. The polymer printing research at ORNL relied on a legacy manufacturer to generate the g-code that was used to print objects. This wasn't an ideal solution for our project because we had no control of what the machine was being commanded to do. We did our best to make the available solutions work for the car, but the STL file was more than a million triangular faces and resulted in more than a million lines of g-code. While the current solutions worked for small parts, a large object like the car caused most programs to freeze or crash. Sometimes, simply loading the object failed. We knew we needed to develop our own solution, and so we started development on what would ultimately become the ORNL Slicer.

Now, nearly a decade later, the development of ORNL Slicer is still very active. We have a team of engineers dedicated to the research and development efforts to advance the state of the art of slicing and toolpath generation software. Those efforts provide the framework for the knowledge and lessons learned in this book.

The idea for this book started with a journal publication that provided a high-level overview of the entire slicing process. To the best of our knowledge, such a publication does not exist. As researchers, we are trying to constantly push the envelope with slicer development and publish as much of our work as we can. For every publication, we struggle with citations about relevant work to reference for slicing. We set out to publish an article that detailed the entire slicing process, something most daily slicing users don't fully understand, and to develop a portfolio about the basics of slicing. Though we didn't invent the slicing process, we wanted to document it from our perspective to educate others. This first publication grew a bit longer than anticipated, but we thought it to be a robust article worthy of submission, so we submitted it.

The article got rejected. Multiple times.

Several reviewers said the article should be a book, or at least a chapter of a book. Creating a book, or finding someone else's book to submit a chapter to, seemed a daunting task. So, we shelved the paper and continued with our research. One day, while he was visiting ORNL, I spoke with Eric MacDonald about various publications and his role as an editor for different journals. The slicing paper was discussed, and he immediately encouraged me to pursue writing my own book. He opened the conversation with Elsevier, who ultimately accepted our proposal to publish this book.

In the ensuing year, I was able to put together a team (including Eric) and an outline, and together we were able to write this book. For all of us, except Eric, it was our first time writing and publishing a book. The process was more laborious than we could have ever imagined, but it has been well worth it to bring light, and documentation, to the slicing process that so many of us rely on day in and day out. Many users may think of the slicing process as a solved problem, but slicing software has a long way to go and can be a limiting factor for new systems development in additive manufacturing. We hope that this book not only educates users and serves as a reference but also encourages aspiring minds to contribute to the development of advanced slicing and toolpathing solutions through open-source projects.

Oh yeah, and we did print that car. The entire print took just 48 h, and we did it in front of a crowd of over a hundred thousand people. It was a wild success and inspired a lot of additional research—research that is still ongoing to this day. That project would have failed without the ORNL Slicer, which proved the need for ORNL to develop slicing software. We've never stopped developing our slicing software and continue to push the edge of state-of-the-art toolpathing, providing vital support to the innovation that is a trademark of additive manufacturing.

Alex Roschli

CHAPTER

1

Introduction to additive manufacturing

Eric MacDonald[a,b], *Abby Barnes*[a], *Alex Roschli*[a], *Thomas Feldhausen*[a], *Michael Borish*[a], *and Peter Wang*[a]

[a]**Manufacturing Demonstration Facility, Oak Ridge National Laboratory, Knoxville, TN, United States,** [b]**University of Texas at El Paso, El Paso, TX, United States**

Abstract

Additive manufacturing (AM) is the process of constructing an object by adding layer upon layer of material in specific locations, to build the object into its final shape. There are many different AM processes that can be used to add layers of material, such as extrusion, material jetting, and sheet lamination. All of the AM processes are a form of digital fabrication because the object to be constructed must first be modeled with a computer. The modeled object must be converted into machine instructions that an AM system can use to construct the object. This chapter serves as an introduction to the many different AM processes. Subsequent chapters will explore the intricacies of creating machine instructions for the AM processes introduced here.

The history of additive manufacturing

Additive manufacturing (AM), colloquially known as 3D printing, is transforming product fabrication by providing a layer-by-layer method of construction. This layer-by-layer approach offers several unique capabilities that are not possible when using the more traditional, subtractive construction methods, such as machining. Subtractive processes have limitations on the features that are possible to manufacture because of their outside-in approach to construction, that is, removing material instead of adding it. They also require a solid piece of material that completely encompasses the end product, much of which gets removed

https://doi.org/10.1016/B978-0-443-15286-3.00003-5

and thrown away as waste. The additive approach to manufacturing produces less material waste, allows a wider variety of materials that can be used in construction, and enables multimaterial and even graded material transitions within a single part. The layer-by-layer approach also allows the construction of features that are impossible to manufacture when limited to removing material, such as cavities within a solid exterior surface.

The first additive, layer-by-layer construction process was introduced in the 1980s as "rapid prototyping." It was developed as a technology to fabricate low-quality prototypes to evaluate shape (form and fit) without function [1–4]. As the technology grew and improved over proceeding decades, the quality of the fabricated objects also increased in terms of spatial resolution, material options, and performance (i.e., mechanical, thermal, isotropic, chemical, wear, and functionality). This increased quality led to a transition from low-quality prototypes to high-quality, end-use products—a transition that AM is still in the processing of making.

The motivation for increased quality was in part due to the significant advantages that the additive approach has for many industries, including mold manufacturing, customized consumer electronics, and athletic wear, but particularly for the aerospace and biomedical industries. Aerospace applications especially benefit from the ability to design and manufacture a structure that is sufficiently strong, but with the absolute minimum weight, based on topology optimization and use of spatially varying lattices. Furthermore, AM enables the construction of qualifiable, geometrically complex parts, from new, high-temperature super alloys, making them capable of surviving extreme conditions, such as those within a jet engine.

In biomedical applications, new, organic designs can be explored that are inspired by nature, conform to human anatomy, and offer more successful biomedical implants, prosthetics, and orthotics. Biomedical applications also benefit from the strength-to-weight tradeoffs possible with AM, as well as the ability to tailor the Young's modulus of certain materials. Furthermore, by leveraging X-ray computer tomography, structural geometries can be guided by the unique human body of a specific patient, and surfaces can be ideally textured to promote osteointegration.

To highlight the potential for these end-use products, the term "additive manufacturing" arose from the field of rapid prototyping. The suite of layer-based manufacturing technologies that has since been developed is now the target of standardization. The technology is referred to as "3D-printing," which provides a more intuitive understanding that a digital model can be realized into a mechanical geometry by adding material, instead of subtracting it. The collection of technologies has matured, and the manufacturing has expanded into metals and ceramics. Consequently, the application space has dramatically increased, as illustrated by the examples depicted in Fig. 1.1 [5–8].

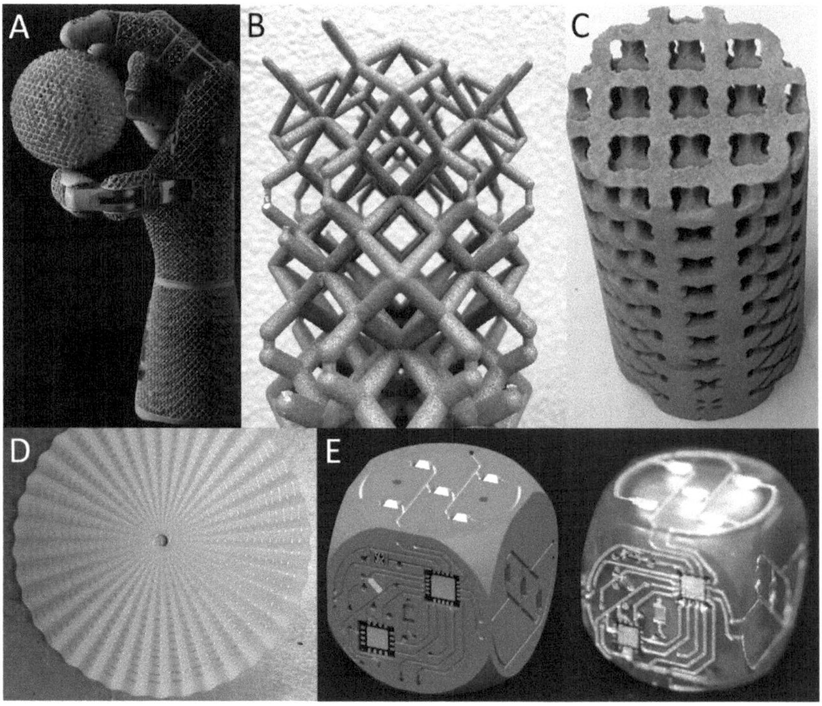

FIG. 1.1 (A) a pneumatic-actuated, 3D-printed hand (Courtesy, Manufacturing Demonstration Facility, Oak Ridge National Laboratory, Department of Energy) [5]; (B) a metallic lattice with spatially-varying struts that manage impact energy; (C) a 3D-printed sand mold that provides complex cavities to enhance metal casting [6], (D) a zirconia sinusoidal disk with a 3D-printed silver spiral dipole antenna [7]; (E) a CAD depiction (left) and an actual 3D-printed version (right) of a gaming die with embedded accelerometer [8].

Taxonomy of additive manufacturing processes

The diversity of AM processes helps demonstrate how AM is transforming manufacturing. The ISO/ASTM 52900 international standards committee defined a taxonomy of AM processes that includes seven broad process categories, which serves as a good framework for exploring the similarities and the unique aspects of slicing for each process. By first understanding the seven process categories, the advantages and challenges of each process can be more easily translated into specific slicing operations. The seven categories are as follows and will be defined in the following subsections:

- Material Extrusion
- Vat Photo Polymerization
- Powder Bed Fusion

- Binder Jetting
- Material Jetting
- Directed Energy Deposition
- Sheet Lamination

Material extrusion

Material Extrusion is the most popular and accessible process category. The original extrusion process was invented in the 1980s by Stratasys, but since then, many of the patents protecting the process details have expired, allowing extrusion systems to be democratized. Now, a wide range of pricing options are available, from hundreds of US dollars for a simple desktop printer, to hundreds of thousands of dollars for large, highly precise production systems with heated build volumes.

In material extrusion, an extruder is translated in 3D space with three to six axes of motion, using either a gantry or robotic system. Naturally, motion control has been inherited from the traditional machining industry, with g-code that describes motion in three dimensions in concert with on/off timing of one or more extrusion tools. Fig. 1.2A illustrates one of the most popular desktop systems, while Fig. 1.2B is an example printed chair made from a large-format material extrusion system [9].

The process works by selectively depositing material through an extrusion nozzle, using either filament or pellet feedstock. This process commonly uses engineering-grade thermoplastics that provide sufficient

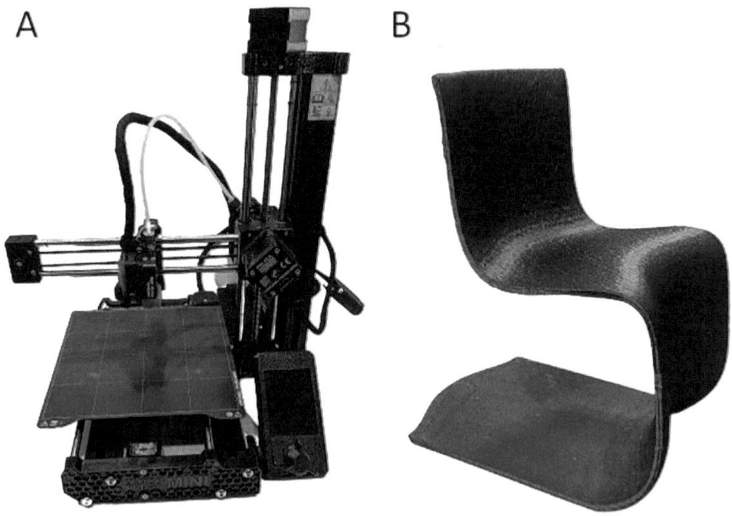

FIG. 1.2 (A) a popular brand of material extrusion desktop printer; (B) an example printed structure.

robustness to withstand many harsh environments (e.g., chemical, mechanical, or temperature exposure). Several thermoplastics are commercially available and are qualified for both biomedical and aerospace applications (e.g., ULTEM™, PEEK and PEKK). Additionally, additives that further improved material properties have been recently demonstrated in thermoplastic feedstock [10].

Vat photopolymerization

Vat photopolymerization features a vat of liquid, photo-curable polymer whose surface is selectively cured with focused light, such as a laser beam or a projection system [11,12]. The process operates in ambient conditions and has high resolution due to the natural advantages of optics, as well as smooth surface finish because the feedstock material begins as a liquid. Depending on the energy source, slicing can vary dramatically for the different versions of this process category. In the case of a laser, a path must be created for tracing the laser across the liquid surface with motor-controlled mirrors. For projection systems, each layer of the fabrication is captured as a bitmap which controls an array of micro mirrors to direct light energy to produce a binary or gray-scale 2D image on the surface of the liquid polymer. Fig. 1.3 shows an example of a vat photopolymerization fabrication.

Two-photon vat photopolymerization is a specialized volume-based version of vat photopolymerization that uses a beam of photons to cure any location within the three-dimensional volume, instead of just the

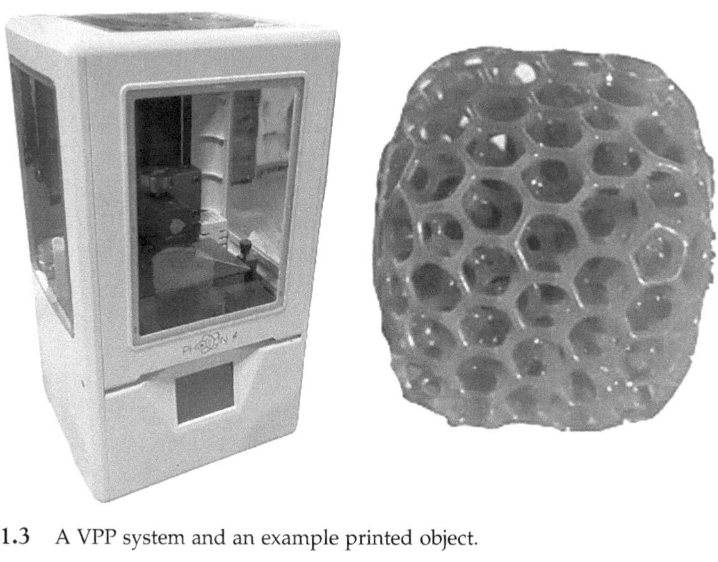

FIG. 1.3 A VPP system and an example printed object.

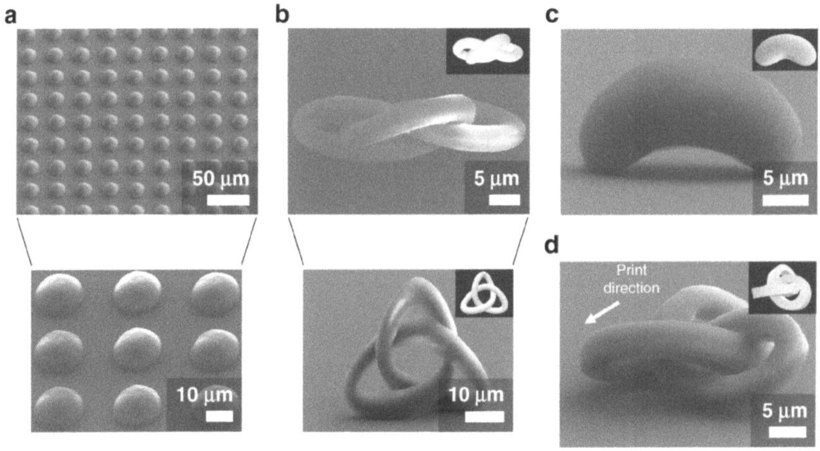

FIG. 1.4 Example prints with two-photon vat photopolymerization. *From P. Somers, et al., Rapid, continuous projection multi-photon 3D printing enabled by spatiotemporal focusing of femtosecond pulses, Light Sci. Appl. 10 (2021) 199.*

surface, of the vat of photocurable polymer. and. This unique process provides the highest level of spatial resolution (approximately 100 nm) and uses a Galvo scanner to move a laser beam and cure the resin within 3D space. Fig. 1.4 depicts examples of the high-resolution printing possible with two-photon printers [13].

Powder bed fusion

Powder bed fusion is similar to vat photopolymerization; it involves selectively melting areas of a layer of powder (typically less than 100 μm thick) using an energy source such as a laser or electron beam. Feedstocks include a wide range of available alloys, and polymer feedstock is available for laser powder bed systems with heated build volumes. The resulting polymer structures are among the strongest of their class. The powder bed fusion process can produce intricate geometries with nearly full density, which is why it is the most broadly adapted AM process in industrial settings. Currently, it is used to manufacture next-generation jet engine components as well as biomedical implants.

Using either a laser that is controlled by mirrors, or an electron beam that is electromagnetically controlled, path planning generally involves the creation of scan paths for each layer. The directed energy of an electron beam is usually much more powerful and heats much more rapidly than that of a laser. Additionally, the motion of the electron beam can be altered rapidly, introducing the possibility of novel paths that quickly and

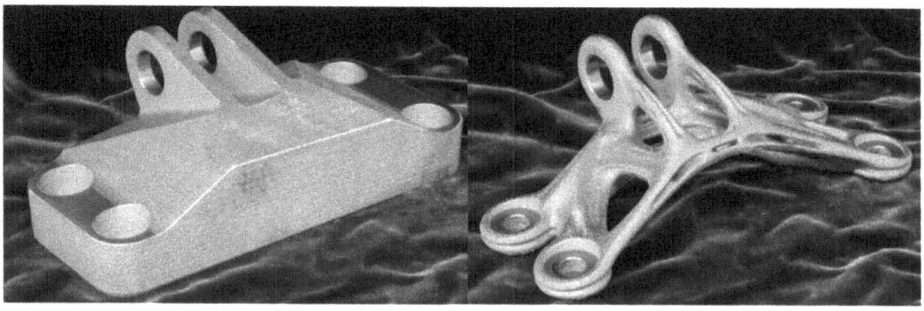

FIG. 1.5 General Electric's Bracket Challenge: the traditional bracket design (left) and a winning design constructed using powder bed fusion (right). *From W.T. Carter, et al., The GE aircraft engine bracket challenge: an experiment in crowdsourcing for mechanical design concepts. 2014 International Solid Freeform Fabrication Symposium. University of Texas at Austin, 2014.*

frequently jump to disparate regions of the powder bed. This method can be exploited to tailor the thermal gradients and histories, ultimately producing different microstructures. Consequently, the scan pathing of electron beam melting is an area of active research. Fig. 1.5 illustrates the General Electric Jet Engine Bracket Challenge in which the design and process freedom of powder bed fusion was leveraged to create a bracket that met all structural requirements while requiring the minimum weight—crucial to fuel economy in aviation [14].

Binder jetting

Binder jetting also uses a powder bed but instead of melting material, ink-jetting technology is used to spray a binder onto the appropriate areas of the powder surface [15]. After binder is deposited on a layer, additional powder is dispensed with a rake or roller, and the process repeats to eventually complete a structure with many layers. Usually, the binder can be deposited in multiple areas simultaneously because of the nature of ink-jetting. Therefore, instead of using g-code, the pathing for each layer is a binary mask that is captured in a CLI file and converted to a black-and-white bitmap for printing. CLI files efficiently describe binary images by listing the points of a contour—the places where regions of black and white meet. However, there are research opportunities involving the use of gray-scale print patterns, which would allow more or less binder to be deposited in specific areas. This could facilitate the removal of mold subsections or modulate the density of the final part.

Generally, the fabricated structure is a "green" part that needs to be sintered in a furnace. However, at least in the case of Furan-based sand printing for metal castings, the printed part is final and serves as a disposable single-use mold. In the case of printed metal and ceramic structures, green parts are processed with additional thermal cycles to remove binder and sinter particles while maintaining the original geometry.

Material jetting

Like binder jetting, material jetting uses ink-jetting technology, but instead of depositing a binding material, the build material is deposited directly and cures immediately prior to the application of additional layers. Material jetting can print multiple materials simultaneously to vary the stiffness or color, as well as to deposit both build and sacrificial support material in a single layer. Material jetting nozzles are usually parallel in a line, so deposition is completed line-wise which provides faster production rates than material extrusion.

The original material jetting systems used photocurable resins with low viscosity, which is required for ink-jetting, followed by a broad UV lamp for curing. However, more recent production systems can now inkjet nanoparticle solutions of either ceramics or metals, which dry in-situ and are then sintered after printing. Fig. 1.6 depicts the XJET system, which jets nanoparticles and is capable of printing ceramics with intricate detail [7].

Directed energy deposition

Directed energy deposition (DED) uses a heat source (e.g., laser, electron beam or arc) to melt a powder or wire feedstock and deposit the melted material in a specific location. Generally, DED is confined to metal feedstock but ceramic and polymers have reportedly been used as well [16]. The major advantage of DED is the rapid deposition rate, which can be orders of magnitude larger than other AM processes but come at the expense of spatial resolution. Within DED processes, wire-based

FIG. 1.6 XJET 1400 Carmel 3D Printer and example zirconia part. *From Y. Oh, et al., Microwave dielectric properties of zirconia fabricated using NanoParticle Jetting™, Addit. Manuf. 27 (2019) 586–594.*

systems generally have higher deposition rates while blown-powder systems have higher resolution. Additionally, blown-powder DED systems are capable of blending various powders to manufacture objects with functionally graded materials.

Recently, there has been a push to integrate various DED processes within CNC machining systems. This integration, often called hybrid manufacturing, captures the benefits of both machining and AM by compensating for their associated disadvantages [17]. Such hybrid systems are capable of the repeatability, dimensional accuracy, and superior surface finish associated with machining, but also have less material waste, the geometrical freedom, and the graded materials that are possible with AM.

Sheet lamination

Sheet lamination uses layers of sheets of paper, polymer, or metal. Each sheet is bonded to the previous layer and then machined to the correct geometry. The sheets are bonded with either an adhesive or another bonding technique, such ultrasonic consolidation. Liquid adhesives can be used to bond paper or polymer sheets, while ultrasonic consolidation uses a sonotrode to bond layers of metal by simultaneous application of pressure and 40,000 Hz vibrations. Ultrasonic consolidation is one of the lowest-temperature metal AM processes and consequently can be used to insert electronics and sensors that would otherwise not be able to survive manufacturing conditions [18,19].

Software support of additive manufacturing

AM uses a digital paradigm in which geometries are virtually created, on a computer, before being physically constructed. The translation from a virtual geometry to a physical object is complex, often involving multi-axis motion, multiple tool heads, multiple materials, and a relatively large variety of material properties within a single object. As such, there is a continuous redefinition of the possible and most ideal approaches to AM. In turn, the software required to design, model and plan for AM machines and their products is becoming increasingly complex.

Software processing in AM is often called "slicing" and, likewise, the software doing this processing is usually referred to as a "slicer." Slicing involves fitting toolpaths, which direct the route of the tool used during construction, to the layers of the geometry being sliced. Slicing can have a dramatic impact on the structural integrity and surface finish of the final object. This is due to many factors, including: (1) quantization error introduced when digitizing the intended geometry into an approximate model, (2) anisotropy and complex thermal histories introduced by the AM process and its toolpathing, and (3) control over the type, amount, location, connection, and strength of the material being used during construction. Optimization of slicing software is therefore critical to the advancement of AM.

This book supports the growing field of AM slicing software by providing a comprehensive summary of the knowledge and processing required to convert an intended geometry, stored in a digital file, into a list of machine commands that an AM system uses to construct an object. The intended audience of this text is anyone interested in AM that would like to understand, develop, enhance, or even simply use slicing software. The text will provide an understanding of the fundamental algorithms behind slicing, the underlying mathematics, and explore future trends in this software discipline.

Organization of text

In pursuit of an exploration of all the factors and steps involved in the slicing process for any AM system, this book covers many topics, including how a machine's motion informs its toolpaths and geometric limitations, the necessary characteristics of a virtual 3D object, the geometrical logic of creating layers and fitting them with toolpaths, the types of toolpaths and their optimizations for specific types of AM, and the instructions that a 3D printer can read to construct an object. The chapters are organized such that proceeding chapters use the information provided in previous chapters. However, each chapter can also stand on its own within the scope of its titular topic.

In the next chapter, this book will provide a brief overview of the mathematics that describe the motion of AM systems, which is fundamental to toolpath planning. Using the defined mathematical terms, *Chapter 3: Motion of platforms* will then survey the different motion platforms and configurations of 3D printing machines, highlighting the advantages and disadvantages of each. These two chapters are meant to provide a basic understanding of the mechanical motion that toolpaths ultimately control.

The discussion of software processing begins in *Chapter 4: Geometry data storage,* which summarizes the steps necessary to convert a 3D model of an object to a 3D mesh. The 3D mesh is necessary for cross-sectioning the object into layers, which is where the term "slicing" comes from and is discussed in *Chapter 5: Cross-sectioning*. *Chapter 6: Closed-loop toolpath generation*, *Chapter 7: Space-filling toolpath generation*, *Chapter 8: Open-loop toolpath generation*, *Chapter 9: Support, raft, brim, and skirt pathing*, *Chapter 10: Path modifiers*, *Chapter 11: Travels, optimizations, and ordering*, *Chapter 12: Pathing considerations for extrusion: Pellet, filament, concrete, and thermoset*, *Chapter 13: Pathing considerations for DED: Arc welding and laser welding*, and *Chapter 14: Planar slicing for nonextrusion AM processes* discuss the process of generating the optimal toolpaths for the layers produced by cross-sectioning. Then, *Chapter 15: Off-axis and nonplanar slicing*, *Chapter 16: Pathing considerations for 5-axis systems*, and *Chapter 17: Pathing considerations for hybrid systems* explore some unique and more advanced pathing strategies, including off-axis, nonplanar, 5-axis, and hybrid toolpathing. The result of path planning, the g-code file, is discussed in depth in *Chapter 18: The g-code file*. The last chapter, *Chapter 19: Closing the loop*, looks at the future of slicing software and discusses how next-generation path planning can be implemented for better process monitoring and control.

References

[1] E. MacDonald, R. Wicker, Multiprocess 3D printing for increasing component functionality, Science 353 (6307) (2016) aaf2093.
[2] I. Gibson, D. Rosen, B. Stucker, Additive Manufacturing Technologies: 3D Printing, Rapid Prototyping, and Direct Digital Manufacturing, second ed., Springer, New York, NY, 2015.
[3] Wohlers Associates, Inc., Wohlers Report 2017: 3D Printing and Additive Manufacturing State of the Industry: Annual Worldwide Progress Report, Wohlers Associates, Incorporated, 2017.
[4] I. Yadroitsev, et al. (Eds.), Fundamentals of Laser Powder Bed Fusion of Metals, Elsevier, 2021.
[5] L.J. Love, R.F. Lind, J.F. Jansen, Mesofluidic actuation for articulated finger and hand prosthetics, in: 2009 IEEE/RSJ International Conference on Intelligent Robots and Systems, IEEE, 2009.
[6] B. Vuksanovich, et al., Non-destructive inspection of sacrificial 3D sand-printed molds with geometrically complex lattice cavities, Int. J. Met. (2022) 1–10.
[7] Y. Oh, et al., Microwave dielectric properties of zirconia fabricated using NanoParticle Jetting™, Addit. Manuf. 27 (2019) 586–594.
[8] E. Macdonald, et al., 3D printing for the rapid prototyping of structural electronics, IEEE Access 2 (2014) 234–242.
[9] L. Love, et al., Commercialization of Big Area Additive Manufacturing, ORNL Report, BT-, vol. ORNL/TM-20, 2020.
[10] L.J. Love, et al., The importance of carbon fiber to polymer additive manufacturing, J. Mater. Res. 29 (17) (2014) 1893–1898.
[11] M. Pagac, et al., A review of vat photopolymerization technology: materials, applications, challenges, and future trends of 3D printing, Polymers 13 (4) (2021) 598.
[12] F. Zhang, et al., The recent development of vat photopolymerization: a review, Addit. Manuf. 48 (2021) 102423.
[13] P. Somers, et al., Rapid, continuous projection multi-photon 3D printing enabled by spatiotemporal focusing of femtosecond pulses, Light Sci. Appl. 10 (2021) 199, https://doi.org/10.1038/s41377-021-00645-z.
[14] W.T. Carter, et al., The GE aircraft engine bracket challenge: an experiment in crowdsourcing for mechanical design concepts, in: 2014 International Solid Freeform Fabrication Symposium, University of Texas at Austin, 2014.
[15] A. Mostafaei, et al., Binder jet 3D printing—process parameters, materials, properties, modeling, and challenges, Prog. Mater. Sci. 119 (2021) 100707.

[16] I. Gibson, et al., Directed energy deposition processes, in: Additive Manufacturing Technologies: 3D Printing, Rapid Prototyping, and Direct Digital Manufacturing, Springer, New York, NY, 2015, pp. 245–268.
[17] T. Feldhausen, et al., Review of computer-aided manufacturing (CAM) strategies for hybrid directed energy deposition, Addit. Manuf. (2022) 102900.
[18] A. Bournias-Varotsis, et al., Ultrasonic additive manufacturing using feedstock with build-in circuitry for 3D metal embedded electronics, Addit. Manuf. 29 (2019) 100799.
[19] A. Hehr, et al., Integrating fiber optic strain sensors into metal using ultrasonic additive manufacturing, JOM 70 (2018) 315–320.

CHAPTER

2

Mechanical systems and kinematics

Peter Wang[a], *Abby Barnes*[a], *Alex Roschli*[a], *Michael Borish*[a], *Thomas Feldhausen*[a], *and Eric MacDonald*[a,b]

[a]Manufacturing Demonstration Facility, Oak Ridge National Laboratory, Knoxville, TN, United States, [b]University of Texas at El Paso, El Paso, TX, United States

Abstract

Kinematics is the study of how motion is achieved without reference to the forces that create the motion. This includes the basic structures of all robotic systems, such as links and joints, as well as their mathematical representations. This chapter will provide a brief overview of the types of motion used in additive manufacturing (AM) systems, as well as the mathematical equations that govern them. Rotations are explained using both Euler rotations and quaternions. Combined rotation and translations are shown using homogeneous transform matrices. The robot Jacobian, which relates joint velocities to the end effector translational velocities, is explained with an example calculation. This section will give the reader the basic mathematical foundation in kinematics that is necessary to understand the mechanical underpinnings of path planning for AM systems.

Introduction

A machine is a mechanical system that transforms energy into a productive behavior. For example, pedaling a bicycle to light up a light bulb converts the chemical energy of a living being into the physical movement of a mechanical system, then converts that movement into electricity. Another example is an electric motor powering a car, in which electrical energy is converted to kinetic energy (motion) through magnetic forces in the motor and mechanical forces in the drive train. This conversion of electrical energy to motion is the principle that drives additive manufacturing (AM) systems today.

https://doi.org/10.1016/B978-0-443-15286-3.00022-9

The mechanisms through which the forces are transmitted can be combinations of cams, gears, linkages, and other machine elements. These combinations are designed to generate specific profiles of output motion, which in turn generate specific actions or behaviors. The motion of these components, or groups of components, can be directly calculated from the physical geometric constraints between the components. The study of this motion, without consideration for the forces acting on or in the system, is called kinematics. This chapter will first discuss the different types of mechanisms that can be used to generate motion, followed by the mathematics that describe the motion achieved by such mechanisms [1,2].

Links, joints, and mechanisms

To transform energy into motion, machines contain mechanisms that are composed of links and joints. Links are the rigid components that cannot generate motion by themselves, such as gears, cams, and bars. Joints connect links together, allow the links to move, and combine the motion of connected links. To be effective, a mechanism must have at least one fixed base, the ground link, that does not move and provides leverage to the rest of the system. Additionally, every machine must have an end effector, often the tool that is being manipulated, which is the point that culminates the motion of all the mechanisms and completes the action that the machine is designed to do. Using this framework, the kinematics of machines and mechanical systems can be fully described and analyzed.

The joints of a mechanism typically provide the mechanism's degree(s) of freedom (DOF), which are the number of distinct motions a mechanism can perform along any given axis. Joints that provide one DOF are called full joints, while joints that provide two DOF are called half joints. For example, the joints between gears and cams are considered half joints, because they can provide two distinct motions. Gears can provide rotation around two different axes, while cams can provide either rotation and translation along one axis, or rotation around one axis and translation along a different axis.

Conversely, the joints on bars and other rigid bodies are generally composed of half joints that are either revolute, providing rotation, or prismatic, providing translation. However, revolute and prismatic joints can be combined to create joints with more DOF, such as universal, cylindrical, and spherical joints, which are illustrated in Fig. 2.1. A universal joint is a combination of two orthogonal revolute joints and has 2 DOF, making it a full joint. A cylindrical joint is another full joint formed by the combination of a prismatic and a revolute joint. A spherical joint has three DOF and is formed by the combination of three orthogonal revolute joints. Collectively, these joints are the fundamental points of motion of any mechanical system.

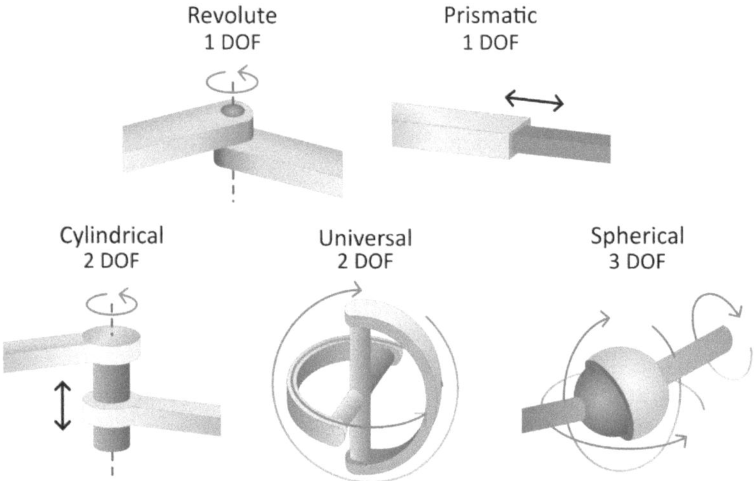

FIG. 2.1 Some of the most common types joints, such as the revolute and prismatic joints, and their corresponding degrees of freedom (DOF). The spatial joints, such as the cylindrical, universal and spherical joints, are also frequently seen.

Note that, although each joint has its own DOF, the DOF of a mechanism is not equal to the simple addition of the DOF of all its joints. Only the distinct motions along or around a distinct axis are included in a mechanism's total DOF. As such, the maximum number of DOF of any mechanism or machine in three-dimensional space is six: three directions of translation (i.e., along the X, Y, and Z axes) and three axes of rotation (i.e., around the X, Y and Z axes).

Mechanisms can be further categorized as either planar, spherical, or spatial, each of which is illustrated in Fig. 2.2. Planar mechanisms work entirely in one plane, while spherical mechanisms move on a sphere. Both of these mechanisms have up to three DOF for their end effector:

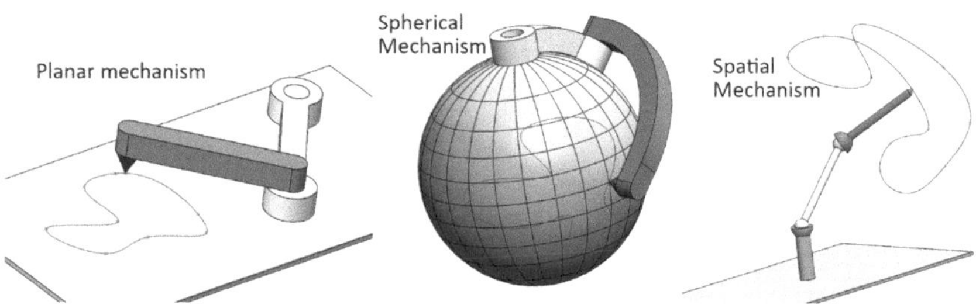

FIG. 2.2 A planar mechanism (left), a spherical mechanism (center), and a spatial mechanism (right), where the planar moves in a two-dimensional plane known as a Euclidian plane, the spherical mechanism moves on a sphere and the spatial mechanism moves through three-dimensional space.

translation along two axes and rotation around one axis. Spatial mechanisms, on the other hand, move through three-dimensional space with translation along all three axes and can include up to three axes of rotation as well. These systems have anywhere between three and six DOF, depending on how many axes of rotation are included. For almost all AM systems, a three-dimensional component is the final desired output, which necessitates the use of spatial mechanisms.

Mechanisms can also be categorized as either serial or parallel. A serial mechanism will have a single chain of links and joints controlling the end effector. A common example of a serial mechanism is a robot arm, in which the base and end effector are connected by a series of links that each have no more than two joints, as shown in Fig. 2.3A. A parallel mechanism, on the other hand, uses multiple serial mechanisms to control a single end effector. A common example of this is the Stewart Platform, which is seen in amusement park rides and is shown in Fig. 2.3B. The platform is positioned by six hydraulic rams with spherical joints at either end. Each hydraulic ram is a serial chain mechanism, and they collectively form a parallel mechanism. There are more complex mechanisms, but they are not commonly seen because they are difficult to design. Therefore, they will not be covered in this book.

As has been implied so far, the axes of motion of a machine are conceptually described in a global coordinate system, such as XYZ, which is independently fixed to a point in space. In other words, it is not affected by any motion of the machine. However, each component of a mechanism or machine also has its own conceptual coordinate system that is rigidly attached to it, such that each component remains fixed within its own coordinate system while the other components move around it. The orientation of these coordinate systems depends on the current orientation of its corresponding mechanism. Any coordinate system, global (also called

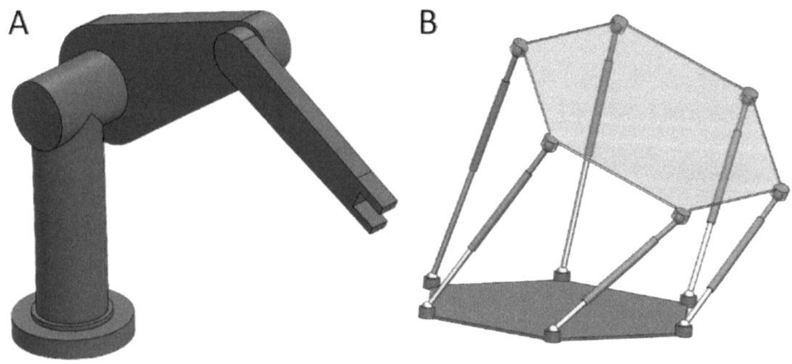

FIG. 2.3 (A) A four degree-of-freedom serial mechanism; (B) a six degree-of-freedom parallel mechanism.

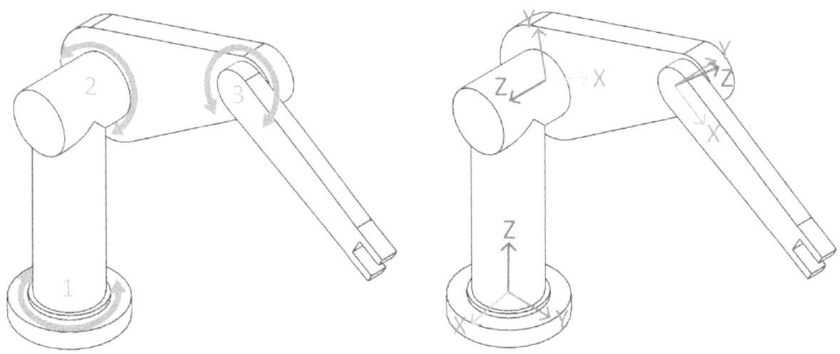

FIG. 2.4 A robotic arm with three revolute joints (left) and the corresponding coordinate frames of each link (right).

world) or local, can be called a "frame of reference" or simply a frame. A mechanical system or machine can thus be modeled as a group of frames whose movements are constrained by geometric relationships. This is the conceptual foundation of kinematics and is illustrated in Fig. 2.4.

Mathematical representations of mechanical motion

This section is meant to serve as a brief overview of the mathematics involved in the kinematic analyses of AM machines. It is important to note that robot kinematics can be calculated in many ways, and the most appropriate way is often determined by the dynamics of the system in question. AM machines are assumed to be rigid-body systems, meaning that components do not deform when an appropriate amount of force is applied. As such, this chapter will not discuss the kinematics of fluid, highly elastic, or plastic systems, because they are irrelevant in the context of AM. Additionally, a complete discussion of the mathematics of rigid-body systems is outside the scope of this book. The mathematical foundations of rigid-body dynamics, such as Lie groups, will not be explained here. As a result, this section should not require such an in-depth understanding of the underlying mathematics to gain a basic understanding of the kinematic analyses relevant to AM.

Degrees of freedom

As previously mentioned, every machine has a number of degree(s) of freedom (DOF), which are the number of distinct motions it can perform along or around any given axis. The DOF of a robotic system are a key metric by which its performance is measured. Recall that, in three-dimensional space, there are a maximum of three translational and three rotational DOF—a total of six DOF. Depending on the kinematic configuration, the end effector of a robotic system may have all six DOF, or it may have fewer.

The total DOF, sometimes referred to as mobility, of a mechanical system is calculated using the mobility equation, shown in Eq. (2.1).

$$M = K(N - 1 - j) + \sum_{i=1}^{j} f_i \tag{2.1}$$

In this equation, M is the total degrees of freedom of the mechanical system, K is the number of parameters needed to specify the position of a single rigid body or link (often the maximum number of DOF, three for two-dimensional space and six for three-dimensional space), N is the total number of links including the ground link, j is the total number of joints, and f_i is the total degrees of freedom for each joint, i.

For example, an ABB 4600 system works in three-dimensional space, so $K=6$. The number of links in the system, including the ground link, is 7, the total number of joints is 6, and each joint has 1 DOF. Substituting into Eq. (2.1) results in $M=6$ DOF. On the other hand, a four-bar linkage has four links, four 1 DOF joints and works in two-dimensional space, so $K=3$. Substituting into Eq. (2.1) results in $M=1$ DOF.

Understanding the DOF of a robotic system, as well as the DOF required for any given task, helps a user select an appropriate robotic platform. Due to the majority of AM printers using planar layers, only three DOF are usually needed. However, the mechanical or robotic system may have more than three DOF, which enables the system to approach the print from different orientations.

Orientation

There are several ways to represent an object's position within a three-dimensional Euclidean space, but the most common method—the one that has been used in this chapter so far—is to use the Cartesian coordinate system and its X, Y, and Z axes. Within this system, the position of a point is simply defined by its X, Y, and Z coordinates. However, when describing the motion of an object, points are not sufficient. Describing how something is moving and where it is going to be, rather than its current position, requires a different kind of information: a vector and its orientation. The orientation of a vector defines the direction an object is moving by describing the conceptual rotations need to reach the direction from a reference direction. The reference direction can be an axis of a frame or some other vector within the frame.

There are, of course, different ways to mathematically describe these conceptual rotations. The common ones are the Euler angle, the angle-axis, and the quaternion. Euler angles describe orientation in terms of three distinct rotations about the axes of a coordinate system. They can use any combination of the X, Y, and Z axes, as long as no axis is consecutively referenced. For example, a ZYZ rotation will reach any target orientation, but ZZY will not. The angle-axis representation, on the other hand, defines a rotation by its angular displacement from a specified vector that is not necessarily a coordinate frame axis. This can be written as the pair (a,θ), where a is a normalized vector that represents the direction of the axis of rotation and the angle θ indicates the magnitude of the rotation. This can be encoded as $n = \theta a$, where the direction of vector n is the axis of rotation, and the magnitude of vector n is the amount rotated about that axis. Quaternions encode the angle-axis information, a and θ, into a four-dimensional vector: $\left(cos\frac{\theta}{2}, a_x sin\frac{\theta}{2}, a_y sin\frac{\theta}{2}, a_z sin\frac{\theta}{2}\right)$. The first element of this vector corresponds to the magnitude of the rotation and is often referred to as the "scalar term." The following three elements refer to the X, Y, and Z elements of the vector that represents the axis of rotation, which are often collectively referred to as the "vector term."

Rotations: Changes in orientation

To rotate an object in space, its orientation needs to be changed. This can be achieved with rotation matrices or quaternion operations. Like Euler angles, rotation matrices describe rotations in terms of the axes of a frame. The rotation matrices for the X, Y, and Z axes are shown in Table 2.1. Rotating around multiple axes is done by treating them as successive rotations. Successive rotation matrices are multiplied together, from left to right, in the order that the rotations will be performed. For example, a ZYX Euler angle is equivalent to the product of a Z-axis, a Y-axis, and an X-axis rotation matrix. Note that, because orientation is defined in terms of rotations, even a single rotation is essentially the combination of two rotations and therefore requires multiplication of the initial direction vector and the rotation matrix.

TABLE 2.1 The rotation matrices used in three dimensional space.

Rotation matrices		
X—Rotation	**Y—Rotation**	**Z—Rotation**
$R_x(\theta) = \begin{bmatrix} 1 & 0 & 0 \\ 0 & \cos\theta & -\sin\theta \\ 0 & \sin\theta & \cos\theta \end{bmatrix}$	$R_y(\theta) = \begin{bmatrix} \cos\theta & 0 & \sin\theta \\ 0 & 1 & 0 \\ -\sin\theta & 0 & \cos\theta \end{bmatrix}$	$R_z(\theta) = \begin{bmatrix} \cos\theta & -\sin\theta & 0 \\ \sin\theta & \cos\theta & 0 \\ 0 & 0 & 1 \end{bmatrix}$

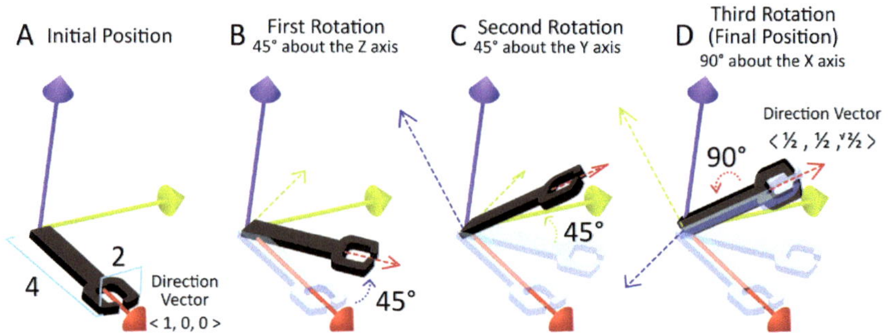

FIG. 2.5 The rotations about the Z, Y and X axes that are needed to change the orientation of a tool with an initial direction vector of $\langle 1 \;\; 0 \;\; 0 \rangle$ and a final direction vector of $\left\langle \frac{1}{2} \;\; \frac{1}{2} \;\; \frac{\sqrt{2}}{2} \right\rangle$. The tool is 4 units long and 2 units wide. For each rotation, the current position of the tool is shown in black, while the previous positions of the tool are shown in light blue. The global coordinate frame is illustrated for spatial reference. Each axis of the local coordinate frame is shown by a dashed line once it diverges from the corresponding global frame axis.

Fig. 2.5 illustrates a ZYX rotation of a tool; the rotation about each axis is shown as a separate step. Note that the rotations occur relative to the local coordinate frame, which means that each rotation changes the orientation of the tool's frame. In the figure, the global frame is illustrated for spatial reference, and each axis of the local frame is indicated by a dashed line once it diverges from the corresponding global frame axis. The initial orientation of the tool is congruent with the positive X axis, defined by the vector $\langle 1 \;\; 0 \;\; 0 \rangle$, and lays flat at 0° relative to the X axis, as shown in Fig. 2.5A. The final orientation of the tool forms a 45° angle with the X, Y and Z axes, defined by the vector $\left\langle \frac{1}{2} \;\; \frac{1}{2} \;\; \frac{\sqrt{2}}{2} \right\rangle$, and lays perpendicular to the X axis, as shown in Fig. 2.5D. The first rotation, shown in Fig. 2.5B, is 45° about the Z axis, so $\theta = \frac{\pi}{4}$. Substituting this value into the Z rotation matrix and then multiplying the matrix by the initial vector yields the rotated direction vector, as shown in Eq. (2.2).

$$R_z\left(\frac{\pi}{4}\right)\langle 1 \;\; 0 \;\; 0 \rangle = \begin{bmatrix} cos\,\frac{\pi}{4} & -\,sin\,\frac{\pi}{4} & 0 \\ sin\,\frac{\pi}{4} & cos\,\frac{\pi}{4} & 0 \\ 0 & 0 & 1 \end{bmatrix} \begin{bmatrix} 1 \\ 0 \\ 0 \end{bmatrix} = \begin{bmatrix} \frac{\sqrt{2}}{2} & -\frac{\sqrt{2}}{2} & 0 \\ \frac{\sqrt{2}}{2} & \frac{\sqrt{2}}{2} & 0 \\ 0 & 0 & 1 \end{bmatrix} \begin{bmatrix} 1 \\ 0 \\ 0 \end{bmatrix}$$
$$= \left\langle \frac{\sqrt{2}}{2} \;\; \frac{\sqrt{2}}{2} \;\; 0 \right\rangle \tag{2.2}$$

To account for the second rotation, shown in Fig. 2.5C, the Z rotation matrix from Eq. (2.2) is multiplied by the Y rotation matrix, as shown in Eq. (2.3).

$$R_z\left(\frac{\pi}{4}\right)R_y\left(-\frac{\pi}{4}\right)\langle 1\ 0\ 0\rangle = \begin{bmatrix} \cos\frac{\pi}{4} & -\sin\frac{\pi}{4} & 0 \\ \sin\frac{\pi}{4} & \cos\frac{\pi}{4} & 0 \\ 0 & 0 & 1 \end{bmatrix} \begin{bmatrix} \cos-\frac{\pi}{4} & 0 & \sin-\frac{\pi}{4} \\ 0 & 1 & 0 \\ -\sin-\frac{\pi}{4} & 0 & \cos-\frac{\pi}{4} \end{bmatrix} \begin{bmatrix} 1 \\ 0 \\ 0 \end{bmatrix}$$

$$= \left\langle \frac{1}{2}\ \frac{1}{2}\ \frac{\sqrt{2}}{2} \right\rangle \tag{2.3}$$

Then, finally, all three rotation matrices are multiplied together, as shown in Eq. (2.4).

$$R_z\left(\frac{\pi}{4}\right)R_y\left(-\frac{\pi}{4}\right)R_x\left(\frac{\pi}{2}\right)\langle 1\ 0\ 0\rangle = \left\langle \frac{1}{2}\ \frac{1}{2}\ \frac{\sqrt{2}}{2} \right\rangle \tag{2.4}$$

Note that the resulting vector is unchanged after the rotation about the X axis. This is because the tool lies on the X axis of its local frame, and the rotations have been done relative to the local frame. The X axis therefore moves with the tool, and a rotation around the X axis does not change the direction of the vector, but it does change the orientation of the object. Evidence of this change in orientation could be found by observing a point in the tool frame that is not on the X-axis.

This can be demonstrated by looking at a corner, rather than the center, of the end of the tool. The end of the tool is a jaw, and for simplicity, the location of the left corner of the jaw will be defined as (4, 1, 0) within the local frame. Eq. (2.5) shows the position vector of the corner of the jaw after just the Z and Y rotations, while Eq. (2.6) shows the position vector of the corner after all the rotations, including the X rotation.

$$R_z\left(\frac{\pi}{4}\right)R_y\left(-\frac{\pi}{4}\right)\langle 4\ 1\ 0\rangle = \left\langle 2-\frac{\sqrt{2}}{2}\quad 2+\frac{\sqrt{2}}{2}\quad 2\sqrt{2} \right\rangle \tag{2.5}$$

$$R_z\left(\frac{\pi}{4}\right)R_y\left(-\frac{\pi}{4}\right)R_x\left(\frac{\pi}{2}\right)\langle 4\ 1\ 0\rangle = \left\langle \frac{3}{2}\ \ \frac{3}{2}\ \ \frac{5\sqrt{2}}{2} \right\rangle \tag{2.6}$$

Now, there is a difference between the final vectors calculated by each equation.

Rotations using quaternions

Quaternions can be used to represent rotations about axes that are not part of the coordinate frame. For example, take the two direction vectors from Fig. 2.5, $\langle 1\ \ 0\ \ 0 \rangle$ and $\left\langle \frac{1}{2}\ \ \frac{1}{2}\ \ \frac{\sqrt{2}}{2} \right\rangle$. The rotation from the first vector to the second vector can be represented in the form of a quaternion, previously stated as $\left(\cos\frac{\theta}{2}, a_X \sin\frac{\theta}{2}, a_Y \sin\frac{\theta}{2}, a_Z \sin\frac{\theta}{2}\right)$. This can be written as $\left(\cos\frac{\theta}{2}, n \sin\frac{\theta}{2}\right)$ where n is the axis of rotation and θ is the angle of rotation about that axis. The first step in obtaining the quaternion representation is to find the axis of rotation, n, which is in fact the vector that is orthogonal to both the initial and final direction vectors. This orthogonal vector n can be calculated using Eq. (2.7), where $\vec{v}_1$ is the initial direction vector and $\vec{v}_2$ is the final direction vector.

$$n = \frac{\vec{v}_1 \times \vec{v}_2}{\|\vec{v}_1 \times \vec{v}_2\|} \tag{2.7}$$

The next step is to calculate angle of rotation, θ, which can be done using Eq. (2.8).

$$\theta = tan^{-1}\left(\frac{\|\vec{v}_1 \times \vec{v}_2\|}{\vec{v}_1 \cdot \vec{v}_2}\right) \tag{2.8}$$

Substituting the example values, $v_1 = \langle 1\ \ 0\ \ 0 \rangle$ and $v_2 = \left\langle \frac{1}{2}\ \ \frac{1}{2}\ \ \frac{\sqrt{2}}{2} \right\rangle$, into Eqs. (2.7) and (2.8) yields $n = \left\langle 0 - \sqrt{\frac{2}{3}}\ \ \frac{\sqrt{3}}{3} \right\rangle$ and $\theta = \frac{\pi}{3}$. The quaternion representation of the rotation, Q_1, is then calculated by substituting those values into the quaternion form, $\left(\cos\frac{\theta}{2}, a_X \sin\frac{\theta}{2}, a_Y \sin\frac{\theta}{2}, a_Z \sin\frac{\theta}{2}\right)$, yielding $Q_1 = \left(\frac{\sqrt{3}}{2}, 0, -\frac{\sqrt{6}}{6}, \frac{\sqrt{3}}{6}\right)$.

As was previously stated, quaternions provide a rotation directly from the initial direction vector to the final direction vector, instead of stepping through different rotations about different axes. This is more efficient, but harder to conceptualize. Additionally, there is a difference in the final orientation of the object's local frame when calculated via rotation matrices vs. quaternions. The vectors representing the X, Y, and Z axes of the local frames, called "basis vectors," can each be different. This is because only the direction vectors are aligned, without consideration for alignment of the rotation vectors.

Table 2.2 summarizes the basis vectors for the frames involved in the current example: the world frame, the object's local frame after the second and third rotation matrices, and the object's local frame after one quaternion rotation. The X basis vector is identical for each rotation because the tool is orientated along the positive X axis. The Y and Z basis vectors, however, are different in each case. This is due to the rotation about the

TABLE 2.2 Comparing the basis vectors for the x, y and z directions after they have been rotated either by the 3 step matrix rotation vs the quaternion rotation.

	X basis vector	Y basis vector	Z basis vector
World frame	$\langle 1\ 0\ 0 \rangle$	$\langle 0\ 1\ 0 \rangle$	$\langle 0\ 0\ 1 \rangle$
Second rotation matrix	$\left\langle \frac{1}{2}\ \frac{1}{2}\ \frac{1}{\sqrt{2}} \right\rangle$	$\left\langle -\frac{1}{2}\ \frac{1}{\sqrt{2}}\ 0 \right\rangle$	$\left\langle -\frac{1}{2}\ -\frac{1}{2}\ \frac{1}{\sqrt{2}} \right\rangle$
Third rotation matrix	$\left\langle \frac{1}{2}\ \frac{1}{2}\ \frac{1}{\sqrt{2}} \right\rangle$	$\left\langle -\frac{1}{2}\ -\frac{1}{2}\ \frac{1}{\sqrt{2}} \right\rangle$	$\left\langle \frac{1}{\sqrt{2}}\ -\frac{1}{\sqrt{2}}\ 0 \right\rangle$
One quaternion	$\left\langle \frac{1}{2}\ \frac{1}{2}\ \frac{1}{\sqrt{2}} \right\rangle$	$\left\langle -\frac{1}{2}\ \frac{5}{6}\ -\frac{1}{3\sqrt{2}} \right\rangle$	$\left\langle -\frac{1}{\sqrt{2}}\ -\frac{1}{3\sqrt{2}}\ \frac{2}{3} \right\rangle$

X vector. The third rotation matrix rotates the coordinate frame by 90°, resulting in a sort of "swap" of the Y and Z basis vectors from the second rotation. However, the application of the quaternion rotation results in completely different Y and Z basis vectors because it uses a completely different axis of rotation, i.e., it does not rotate about the aligned axis.

To demonstrate the equivalency of quaternions and rotation matrices, a single quaternion must be calculated for each rotation, similar to what was done with the rotation matrices. However, in this case, the angles and the axes of rotation are already known. The first rotation is about the Z axis, $\langle 0\ \ 0\ \ 1 \rangle$, with a rotation angle of 45°. The second rotation is about the Y axis, $\langle 0\ \ 1\ \ 0 \rangle$, with a rotation angle of −45°. The third rotation is 90° about the X axis, $\langle 1\ \ 0\ \ 0 \rangle$. Substituting into the quaternion equation yields $Q_1 = \left(cos\frac{\pi}{8}, 0, 0, sin\frac{\pi}{8}\right)$, $Q_2 = \left(cos\frac{\pi}{8}, 0, -sin\frac{\pi}{8}, 0\right)$, $Q_3 = \left(\frac{\sqrt{2}}{2}, \frac{\sqrt{2}}{2}, 0, 0\right)$.

To combine two subsequent rotations in quaternion form, the quaternions must be multiplied, just as with rotation matrices. However, quaternion multiplication is noncommutative, meaning that it matters which order the terms are multiplied in, as shown in Eq. (2.9).

$$Q_1 Q_2 \neq Q_2 Q_1 \tag{2.9}$$

To find the product of two quaternions, a unique quaternion multiplication operation must be used. This multiplication operation is shown in Eq. (2.10), where $Q_a = \left(a_0, \vec{a}\right)$, a_0 is the scalar term of the quaternion, and $\vec{a}$ is the vector term of the quaternion, $\vec{a} = \langle a_x\ \ a_y\ \ a_z \rangle$. Likewise, $Q_b = \left(b_0, \vec{b}\right)$ where b_0 is the scalar term of the quaternion and $\vec{b} = \langle b_x\ \ b_y\ \ b_z \rangle$ is the vector term of the quaternion.

$$Q_a Q_b = \left(a_0 b_0 - \vec{a} \cdot \vec{b}, a_0 \vec{b} + b_0 \vec{a} + \vec{a} \times \vec{b}\right) \tag{2.10}$$

The combination of the quaternions, with Q_1 for the first rotation then Q_2 for the second rotation, is therefore multiplied as shown in Eq. (2.11).

$$Q_{full} = Q_1 Q_2 = \left(\frac{2+\sqrt{2}}{4}, \frac{\sqrt{2}-1}{2\sqrt{2}}, -\frac{\sqrt{2}}{4}, \frac{\sqrt{2}}{4}\right) \tag{2.11}$$

The final rotation, Q_3, would be about the X axis, but the initial direction vector is also on the X axis, so the coordinates would not be changed when $Q_3$is applied. This was discussed in the previous section on rotation matrices, so here, only Q_1 and Q_2 are considered.

As with rotation matrices, changing an orientation via quaternion is essentially two sequential rotations, because orientation is defined in terms of rotations. Therefore, to apply a quaternion rotation to a vector, the vector must be converted into quaternion form. To put a three-dimensional vector of Cartesian coordinates into quaternion form, the vector portion of the quaternion is substituted and a zero is used for the scalar term. If $\vec{v} = \langle v_x \;\; v_y \;\; v_z \rangle$, then the quaternion form is $Q_v = (0, v_x, v_y, v_z)$. Then, Eq. (2.12) is used to find the final position of the vector after the quaternion rotation.

$$v = qpq^{-1} \tag{2.12}$$

In this equation, v is the final position of the vector in quaternion form, p is the initial position of the vector in quaternion form, q is the quaternion rotation to be applied to the vector and q^{-1} is the conjugate of the quaternion rotation. The conjugate of the quaternion rotation is found by negating the vector terms of the quaternion. If $Q_a = (a_0, a_x, a_y, a_z)$ then $Q_a^{-1} = (a_0, -a_x, -a_y, -a_z)$.

Now the quaternion rotation Q_{full} can be applied to the initial direction vector $\vec{v}_1 = \langle 1 \;\; 0 \;\; 0 \rangle$. The initial direction vector's quaternion representation is $p_{v1} = (0,1,0,0)$. The conjugate of Q_{full} is $Q_{full}^{-1} = \left(\frac{2+\sqrt{2}}{4}, -\frac{\sqrt{2}-1}{2\sqrt{2}}, \frac{\sqrt{2}}{4}, -\frac{\sqrt{2}}{4}\right)$. Using Eq. (2.12), the final direction vector is $p_{v2} = Q_{full} p_{v1} Q_{full}^{-1} = \left(0, \frac{1}{2}, \frac{1}{2}, \frac{\sqrt{2}}{2}\right)$, whose vector term is equal to the final direction vector $\left\langle \frac{1}{2} \;\; \frac{1}{2} \;\; \frac{\sqrt{2}}{2} \right\rangle$. This has clearly given the same result as previously found using the rotation matrices.

To demonstrate use of a Q_3 quaternion, this process can be applied to the left corner of the jaw of the tool, which was defined with the coordinates of $j_1 = \langle 4 \;\; 1 \;\; 0 \rangle$. Since these coordinates do not lie on the X axis, Q_3 affects the final coordinates. Eq. (2.13) shows the calculation used to find the final position vector, j_2, which is the vector term of the quaternion p_{j2}.

$$p_{j2} = Q_{full} p_{j1} Q_{full}^{-1} = \left(0, \frac{3}{2}, \frac{3}{2}, \frac{5\sqrt{2}}{2}\right) \tag{2.13}$$

In the equation, Q_{full} is the quaternion used to make the full rotation of the end effector, $Q_{full} = Q_1Q_2Q_3 = \left(\frac{1}{2}, \frac{\sqrt{2}}{2}, 0, \frac{1}{2}\right)$. The inverse quaternion is $Q_{full}^{-1} = (Q_1Q_2Q_3)^{-1} = Q_3^{-1}Q_2^{-1}Q_1^{-1} = \left(\frac{1}{2}, -\frac{\sqrt{2}}{2}, 0, -\frac{1}{2}\right)$, and p_{j1} is the quaternion representation of the initial direction vector, $p_{j1}=(0,4,1,0)$. The resulting quaternion vector term is $\left\langle \frac{3}{2} \frac{3}{2} \frac{5\sqrt{2}}{2} \right\rangle$, which is the same as the final vector found using the previous two methods. The final vector again matches the vector found using rotation matrices.

Rotations with translations

Rotations are obviously not the only changes produced by motion; things also move forward or backward along straight trajectories. These movements are called translations, and a rotation with a translation can be collectively termed a "transformation." To describe a transformation, homogenous transform matrices or dual quaternions can be used. The homogenous transform matrices incorporate translational movements into rotation matrices as shown in Table 2.3, where θ is the angular rotation about the respective axis and d is the distance translated along the axis. Once again, multiple transformations through space are represented by multiplying them together. This can be used to calculate the end effector position for serial chain robotic arms that have multiple links.

When calculating rotations with translations, it is common to limit definitions of movement to the X and Z axes, according to the Denavit-Hartenberg (DH) convention [3]. This convention does not have to be followed, but it provides a common framework for developing the equations of motion of robots. Using this convention, the angle of rotation around the Z axis is denoted as θ and the displacement along the Z axis is denoted as d, similar to the notations used in the homogenous transforms. However, the angle of rotation around the X axis is denoted as α and the displacement along the X axis is denoted as a.

TABLE 2.3 4×4 homogeneous transform matrices.

X-axis homogenous transform $X(\theta, d)$	Y-axis homogenous transform $Y(\theta, d)$	Z-axis homogenous transform $Z(\theta, d)$
$\begin{bmatrix} 1 & 0 & 0 & d \\ 0 & \cos\theta & -\sin\theta & 0 \\ 0 & \sin\theta & \cos\theta & 0 \\ 0 & 0 & 0 & 1 \end{bmatrix}$	$\begin{bmatrix} \cos\theta & 0 & \sin\theta & 0 \\ 0 & 1 & 0 & d \\ -\sin\theta & 0 & \cos\theta & 0 \\ 0 & 0 & 0 & 1 \end{bmatrix}$	$\begin{bmatrix} \cos\theta & -\sin\theta & 0 & 0 \\ \sin\theta & \cos\theta & 0 & 0 \\ 0 & 0 & 1 & d \\ 0 & 0 & 0 & 1 \end{bmatrix}$

When using the DH convention, the Z axis of a joint is defined as the axis of rotation (for a revolute joint) or direction of slide (for a prismatic joint). Generally, the X axis follows the link to the next joint; it is defined by a vector that is parallel to the common normal between the Z axis of the current joint and the Z axis of the next joint in the chain. The direction vector of the common normal can be found by taking the cross product of the direction vectors of the two Z axes. Fig. 2.6 illustrates the DH parameters of a serial chain robotic arm: α represents rotations around the X axis, a represents the displacement along the X axis, θ represents rotations around the Z axis, and d represents displacement along the Z axis.

Table 2.4 shows the DH parameters listed in table form. In this table, i indicates the joint number, and each row shows the rotation and displacement values (α, a, θ and d) that are needed to calculate the position and orientation of the corresponding joint. In a serial chain robot, the position and orientation of a joint are dependent on its relationship to and displacement from the previous joint. Remember that, according to the DH convention, the X axis is the axis that relates the current joint to the next joint in the chain. Therefore, in a DH table, the X axis rotation and

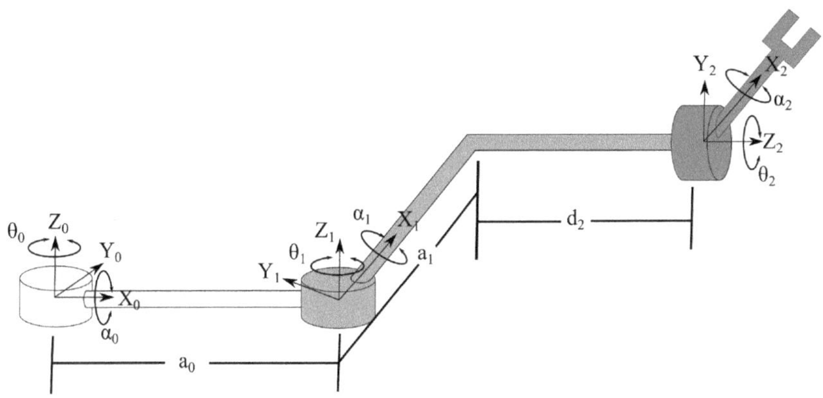

FIG. 2.6 A robot with three revolute joints. The DH values of d_i, θ_i, α_i, and a_i are dimensioned in the figure above.

TABLE 2.4 The Denavit-Hartenberg parameter table for Fig. 2.7.

i	α_{i-1}	a_{i-1}	θ_i	d_i
0	0	0	θ_0	0
1	0	a_0	θ_1	0
2	$\frac{\pi}{2}$	a_1	θ_2	d_2

displacement (α and *a*, respectively) for any given joint are defined in terms of the previous joint, i.e., $i-1$.

A DH table provides an easy way to identify the kinematics of a machine, because the α, *a*, θ and *d* values of each joint can be used to determine what type of joint it is as well as its relationship to other joints. For each joint, the number of parameters that are variables, as opposed to constants, corresponds to the number of DOF that the joint has. According to the DH convention, revolute joints always rotate about the Z axis and prismatic joints will always slide along the Z axis. As such, a revolute joint has a variable θ and a constant *d*, reflecting its single DOF. All the joints in our example (Fig. 2.6) are revolute, so in Table 2.4, every θ is a variable but the other terms are constants. A prismatic joint, on the other hand, would have a constant θ and a variable *d*.

Joints that are made of multiple, simpler joints list each joint separately in a DH table. For example, a universal joint, which is made up of two revolute joints, would list one revolute joint with a constant α (X-axis rotation) of zero and one revolute joint with a constant α of $\frac{\pi}{2}$, while both joints would have a variable θ, as shown in Table 2.5. The $\frac{\pi}{2}$ indicates that the two joints are perpendicular to each other and that the output of the universal joint is dependent on the rotational position of both joints. A spherical joint is similarly represented, which is also shown in Table 2.5. The vector of a universal joint can intersect any point on a sphere, but there is no rotation about the axis of the vector. On the other hand, the vector of a spherical joint can intersect any point on a sphere, and there is rotation about the axis of the vector.

Ultimately, the DH parameters describe joint positions and link geometries. They can be used to calculate the position and orientation of the end effector of a robot in a process that is called "forward kinematics" or a

TABLE 2.5 Representation of complex joints in DH tables.

Joint type	i	α_{i-1}	a_{i-1}	θ_i	d_i
Universal joint	0	0	0	θ_0	0
	1	$\frac{\pi}{2}$	0	θ_1	0
Spherical joint	0	0	0	θ_0	0
	1	$-\frac{\pi}{2}$	0	θ_1	0
	2	$\frac{\pi}{2}$	0	θ_2	0

"kinematic analysis." A kinematic analysis is demonstrated below, in an example analysis of the robotic arm illustrated in Fig. 2.6.

Example kinematic analysis

The forward kinematics of a robot are essentially the product of all the transformations of the joints of the robot. Therefore, the first step in a kinematic analysis is to find the transformations (T) of each joint. The equations defining the transformations are called kinematic equations. The kinematic equation of the first joint, T_0, is determined by the displacement along and rotation about the Z axis. When using homogenous transform matrices, T_0 is equivalent to the Z homogeneous transform matrix of the first joint, as shown in Eq. (2.14). Note that an abbreviated form of the homogenous transform matrix is used, which represents the matrix as a function.

$$T_0 = Z(\theta_0, d_0) \tag{2.14}$$

The kinematic equation for the second joint is a displacement from the first joint that is described by the X homogeneous transform matrix of the first joint and the Z homogeneous transform matrix of the second joint, as shown in Eq. (2.15).

$$T_1 = Z(\theta_0, d_0)X(\alpha_0, a_0)Z(\theta_1, d_1) \tag{2.15}$$

To determine the kinematic equations for the ith joint, continue to multiply by X and Z homogenous transformations, as shown in Eq. (2.16). If the end effector or tool does not lie on the final joint, then an additional set of transformations is required to determine the location of the tool frame.

$$T_i = Z(\theta_0, d_0)X(\alpha_0, a_0)Z(\theta_1, d_1)\ldots X(\alpha_{i-1}, a_{i-1})Z(\theta_i, d_i) \tag{2.16}$$

In the three revolute joint linkage shown in Fig. 2.6, the position of the end effector is calculated by substituting the values from Table 2.4 into the matrices from Table 2.3, then multiplying these matrices together, as shown in Eq. (2.17).

$$\begin{aligned} F_{world} &= Z(\theta_0, d_0)X(\alpha_0, a_0)Z(\theta_1, d_1)X(\alpha_1, a_1)Z(\theta_2, d_2)F_{end} \\ &= Z(\theta_0, 0)X(0, a_0)Z(\theta_1, 0)X\left(\frac{\pi}{2}, a_1\right)Z(\theta_2, d_2)F_{end} \end{aligned} \tag{2.17}$$

In this equation, F_{end} is the vector containing the end effector's coordinates within the last joint's local frame, while F_{world} is the end effector's coordinates within the world frame. Please note that the joints are labeled as follows: the first joint is joint 0, the second joint is joint 1, and the third and last joint is joint 2. F_{end} is expressed in homogeneous coordinates, i.e., perspective coordinates, which are written as $(x, y, z, 1)$. This form allows (x, y, z) coordinates to be multiplied by the homogenous transform matrices.

Substituting numerical values into Eq. (2.16) will find the end effector's coordinates in the global frame. For this example, arbitrary values will be used: $a_0 = 1$, $a_1 = 0.75$, $d_2 = 1.5$ and $F_{end} = (1, 0, 0, 1)$. Eq. (2.17) shows the values substituted into the abbreviated forms of the homogenous transform matrices, while Eq. (2.18) shows an expanded form with the relevant trigonometric functions.

$$F_{world} = Z(\theta_0, 0)X(0,1)Z(\theta_1, 0)X\left(\frac{\pi}{2}, 0.75\right)Z(\theta_2, 1.5)(1,0,0,1)^{\mathrm{T}} \tag{2.18}$$

$$\begin{aligned} F_{world}(x,y,z1) = \Big(& sin\ \theta_0(1.5\ cos\ \theta_1 - sin\ \theta_1(0.75 + cos\ \theta_2)) \\ & + cos\ \theta_0(1 + cos\ \theta_1(0.75 + cos\ \theta_2) + 1.5\ sin\ \theta_1), \\ & cos\ \theta_0(-1.5\ cos\ \theta_1 + sin\ \theta_1(0.75 + cos\ \theta_2)) \\ & + sin\ \theta_0(1 + cos\ \theta_1(0.75 + cos\ \theta_2) + 1.5\ sin\ \theta_1),\ sin\ \theta_2, 1\Big) \end{aligned} \tag{2.19}$$

From this expansion, the values of x, y and z can be found as:

$$\begin{aligned} x = & sin\ \theta_0\ (1.5\ cos\ \theta_1 - sin\ \theta_1\ (0.75 + cos\ \theta_2)) \\ & + cos\ \theta_0(1 + cos\ \theta_1(0.75 + cos\ \theta_2) + 1.5\ sin\ \theta_1) \end{aligned} \tag{2.20}$$

$$\begin{aligned} y = & cos\ \theta_0\ (-1.5\ cos\ \theta_1 + sin\ \theta_1\ (0.75 + cos\ \theta_2)) \\ & + sin\ \theta_0(1 + cos\ \theta_1(0.75 + cos\ \theta_2) + 1.5\ sin\ \theta_1) \end{aligned} \tag{2.21}$$

$$z = \sin\theta_2 \tag{2.22}$$

These are the forward kinematic equations. Using the forward kinematic equations, the final, global frame coordinates (x, y, z) of the end effector can be directly calculated from the current joint positions of the robot. If the angles of rotation of the first, second and third joint are known, then the location of the end effector can be calculated. For example, if $(\theta_0, \theta_1, \theta_2) = (0,0,0)$ then $(x, y, z) = (2.75, -1.5, 0)$ or if $(\theta_0, \theta_1, \theta_2) = \left(\frac{\pi}{6}, \frac{\pi}{3}, \frac{\pi}{4}\right)$ then $(x,y,z) = (2.366, 1.957, 0.707)$. The end effector coordinates can thus be calculated for any possible angle that the joints of the robot can reach.

It is also possible to calculate the inverse of this, in which the desired end effector transformation is known, and the corresponding joint angles are calculated. This is known as inverse kinematics and produces inverse kinematic equations. Inverse kinematics can be solved in a number of ways, but the method described here will continue to use the 4×4 homogeneous transform matrices. Recall that, in these matrices, the last column represents the Cartesian coordinates and the remaining 3×3 matrix represents the rotation matrix that describes the orientation. These components can be labeled $[d]$ and $[R]$, respectively, as illustrated (Eq. 2.23). Any position and orientation, with six degrees of freedom, can be represented in this form.

$$F_{target} = \begin{bmatrix} R_{1,1} & R_{1,2} & R_{1,3} & x \\ R_{2,1} & R_{2,2} & R_{2,3} & y \\ R_{3,1} & R_{3,2} & R_{3,3} & z \\ 0 & 0 & 0 & 1 \end{bmatrix} = \begin{bmatrix} [R] & [d] \\ 0 & 1 \end{bmatrix} \tag{2.23}$$

In this example, F_{end} will be represented as a matrix (F_{endM}) rather than a vector, to keep the transformation in matrix form. From the previous forward kinematics example, the end effector has a translation of 1 along the X axis of the last joint. Since this is only a translational offset, the rotation matrix $[R]$ is a 3×3 identity matrix and $[d]$ is simply the coordinates (1,0,0), as shown in Eq. (2.24).

$$F_{endM} = \begin{bmatrix} 1 & 0 & 0 & 1 \\ 0 & 1 & 0 & 0 \\ 0 & 0 & 1 & 0 \\ 0 & 0 & 0 & 1 \end{bmatrix} \tag{2.24}$$

F_{world} also needs to be in matrix form (F_{worldM}), which is accomplished by substituting F_{endM} for F_{end} in Eq. (2.17), as shown in Eq. (2.25).

$$F_{worldM} = Z(\theta_0, 0)X(0, a_0)Z(\theta_1, 0)X\left(\frac{\pi}{2}, a_1\right)Z(\theta_2, d_2)F_{endM} \tag{2.25}$$

The expression is fundamentally the same as F_{world}, just in matrix form. The definition of F_{worldM} can therefore be found by substituting Eq. (2.20) for x, Eq. (2.21) for y, and Eq. (2.22) for z in Eq. (2.26).

$$F_{worldM} = \begin{bmatrix} R_{1,1} & R_{1,2} & R_{1,3} & x \\ R_{2,1} & R_{2,2} & R_{2,3} & y \\ R_{3,1} & R_{3,2} & R_{3,3} & z \\ 0 & 0 & 0 & 1 \end{bmatrix} \tag{2.26}$$

To begin an inverse kinematic analysis, a point in the global frame must be selected as the target location of the end effector. A point is necessary because there may be an infinite number of solutions if only an orientation is specified—the robot may be able hold an orientation while moving through space. Also, the selected point must be within the workspace of the robot. If the selected coordinate is not within the workspace, then there will be no solutions found. More information on a robot's workspace is detailed in the next section.

To find the joint rotations for the robot in Fig. 2.6 that would place the end effector at the global frame coordinates (1,1,0.5), the first step is to substitute those coordinates into Eq. (2.23), as shown in Eq. (2.27). Note that the [R] component is not relevant here because we are defining location, not orientation. Therefore, the values of the variables in [R] will not be calculated in this analysis. If the robot in question is a spatial mechanism and has more than three degrees of freedom, then orientation angles can also be defined.

$$F_{target} = \begin{bmatrix} R_{1,1} & R_{1,2} & R_{1,3} & 1 \\ R_{2,1} & R_{2,2} & R_{2,3} & 1 \\ R_{3,1} & R_{3,2} & R_{3,3} & 0.5 \\ 0 & 0 & 0 & 1 \end{bmatrix} \tag{2.27}$$

To solve for the inverse kinematics, the current location of the end effector, F_{worldM}, needs to be the same as the target location of the end effector, F_{target}, so the two matrices are set equal to each other.

$$F_{worldM} = F_{target} \tag{2.28}$$

Then, the angles of rotation ($\theta_0, \theta_1, \theta_2$) can be solved for, element by element. They can each be found using only the three defined positions, (x, y, z), because we only have three degrees of freedom. This means the [R] components of the matrices can be ignored. So, the three inverse kinematic equations are:

$$\begin{aligned} x = {} & sin\ \theta_0\,(1.5\ cos\ \theta_1 - sin\ \theta_1\,(0.75 + cos\ \theta_2)) \\ & + cos\ \theta_0(1 + cos\ \theta_1(0.75 + cos\ \theta_2) + 1.5\ sin\ \theta_1) = 1 \end{aligned} \tag{2.29}$$

$$\begin{aligned} y = {} & cos\ \theta_0\,(-1.5\ cos\ \theta_1 + sin\ \theta_1\,(0.75 + cos\ \theta_2)) \\ & + sin\ \theta_0(1 + cos\ \theta_1(0.75 + cos\ \theta_2) + 1.5\ sin\ \theta_1) = 1 \end{aligned} \tag{2.30}$$

$$z = sin\ \theta_2 = 0.5 \tag{2.31}$$

To calculate the angles of rotation, first solve for the value of θ_2. Then substitute this value into the x and y equations and solve for θ_1 and θ_0. There are two valid results which, in radians, are:

$$\begin{matrix} \theta_0 \\ \theta_1 \\ \theta_2 \end{matrix} = \begin{matrix} -1.538 \\ -2.897 \\ 0.5236 \end{matrix} \ \boldsymbol{OR}\ \begin{matrix} 3.0746 \\ -1.889 \\ 0.5236 \end{matrix}$$

The two results show that there are two configurations that the robot can use to reach the desired end effector location. In the first configuration, the elbow will be pointed up, while in the second, configuration the elbow will be pointed down. It is up to the designer or operator to decide which configuration to use. Often, the decision is determined by the presence of outside interferences or by the next movements of the robot. Changing from one configuration to another requires the robot to move through a singularity, which will be explained in a later section. Such configuration changes are not ideal, so maintaining configuration from one movement to the next helps guide which configuration should be chosen for the current movement.

As was mentioned at the beginning of this section, dual quaternions can also be used to calculate transformations. However, as was illustrated in the example of a simple rotation, using dual quaternions requires the manipulation of the properties of quaternions. In dual quaternions, this includes dual numbers. This method is much more complex and so will not be detailed here, but just as with simple rotations, it is ultimately more efficient.

Robot workspaces

The workspace of a robot or mechanism is the full set of positions and orientations that the end effector can reach. Kinematic equations can be used to calculate the workspace of a robot from the geometry of each link and the DOF of each joint, as described in the previous section. Also, as was previously noted, inverse kinematics can only be found for locations that are within a robot's workspace.

The size and shape of a robot's workspace are determining factors in the size and shape of parts it can build. For example, most material extrusion AM systems use a Cartesian robot, which results in a rectangular prism for the workspace, so any shape that fits within this rectangular prism will be reachable by the printhead. However, if the workspace is toroidal, then shapes that must cross the center of the torus will not be printable, leading to this machine being better suited to print cylinders or small parts that fit between the inner and outer radius of the workspace. Fig. 2.7 shows several examples of workspaces of common robotic systems.

If the base of the robot can move, then the workspace can be significantly expanded. The size and of the workspace are no longer limited to that of the robot, but rather by where the robot can be moved. In such cases, the size of a printed part is now limited by a combination of the range of the mobile base and the workspace of the robot. An example of this is shown in Fig. 2.8.

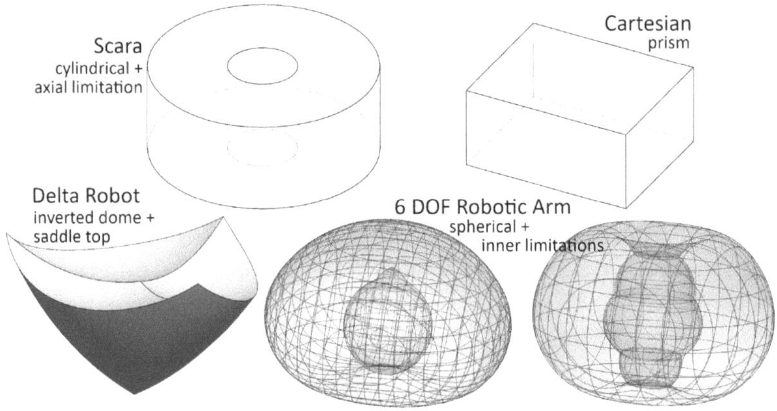

FIG. 2.7 The workspace of various different types of robotic systems. The Scara configuration (top left) has a cylindrical workspace with an axial opening, the Cartesian robot (top right) has a rectangular prism, the delta robot (bottom left) has an inverted dome with a saddle top, and the 6DOF robotic arm has a spherical workspace with inside limitations (bottom center and bottom left).

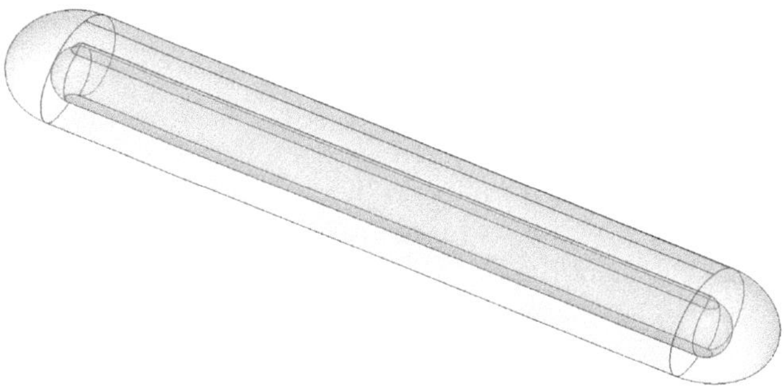

FIG. 2.8 The workspace of a 6DOF robot with a linear track added to increase to total workspace. The round workspace becomes elongated in the direction of the track.

Robot Jacobians

So far, only the location and orientation aspects of motion have been described. One very important factor has been missing: speed. A robot Jacobian establishes the relationship between joint positions, joint speeds, and the end effector's linear and angular velocities. It can also be used to determine the relationship between the joint forces, joint torques, and the end effector's forces and torques.

The robot Jacobian is calculated from the time derivative of the robot's kinematic equations. This can allow a designer to use the torques exerted by the joint motors to understand what is happening at the end effector.

The robot Jacobian can also be inverted to understand the impact that an end effector, such as a heavier print head, has on the joint motors. A robot Jacobian, J, can be calculated using Eq. (2.32), which relates the robot's joint space to its coordinate space.

$$\dot{X} = J\dot{q} \tag{2.32}$$

In this equation, $\dot{X}$ is a 6×1 matrix defining the velocity of the robot's translations and rotations in three-dimensional space. J is the robot Jacobian and is a $6 \times n$ matrix, while $\dot{q}$ is the joint velocities and is an $n \times 1$ matrix, where n is the number of joints in the robot.

The Jacobian can be separated into two distinct parts, $J = \begin{bmatrix} J_v \\ J_\omega \end{bmatrix}$, where J_v is the Jacobian associated with the translational terms, which are equivalent to the partial derivatives of the kinematic equations. J_v is therefore a $3 \times n$ matrix, as shown in Eq. (2.33).

$$J_v = \begin{bmatrix} \frac{\partial x}{\partial q_1} & \frac{\partial x}{\partial q_2} & \cdots & \frac{\partial x}{\partial q_n} \\ \frac{\partial y}{\partial q_1} & \frac{\partial y}{\partial q_2} & \cdots & \frac{\partial y}{\partial q_n} \\ \frac{\partial z}{\partial q_1} & \frac{\partial z}{\partial q_2} & \cdots & \frac{\partial z}{\partial q_n} \end{bmatrix} \tag{2.33}$$

The second part, J_ω, is the Jacobian associated with the rotational terms and is a $3 \times n$ matrix, where n is again the number of joints. This is shown in Eq. (2.34), where ζ is 1 for revolute joints and 0 for prismatic joints, and R_{zn} is the basis vector of the Z axis of each joint n in the robotic system.

$$J_\omega = [\zeta_1 R_{z1} \;\; \zeta_2 R_{z2} \;\; \ldots \;\; \zeta_n R_{zn}] \tag{2.34}$$

ζ is applied to eliminate prismatic joints from the equations of rotation because they do not impart any rotation. The Z basis vector, R_{zn}, for each joint is a 3×1 vector that can be calculated using Eq. (2.35), where R_n is the rotation matrix portion of the homogenous transform matrix from the robot base to the n joint and $\vec{Z}$ is the basis vector for the world Z axis, $\langle 0,0,1 \rangle$.

$$R_{zn} = R_n \vec{Z} = R_n \begin{bmatrix} 0 \\ 0 \\ 1 \end{bmatrix} \tag{2.35}$$

For a simple demonstration of the calculation of a robot Jacobian, take a modified version of the planar robot arm used in the discussion of translations with rotations, shown in Fig. 2.9. This version of the robot arm does not have a bend in the second link, and the tool is no longer mounted at an angle. The tool becomes the end effector of the mechanism. The DH table

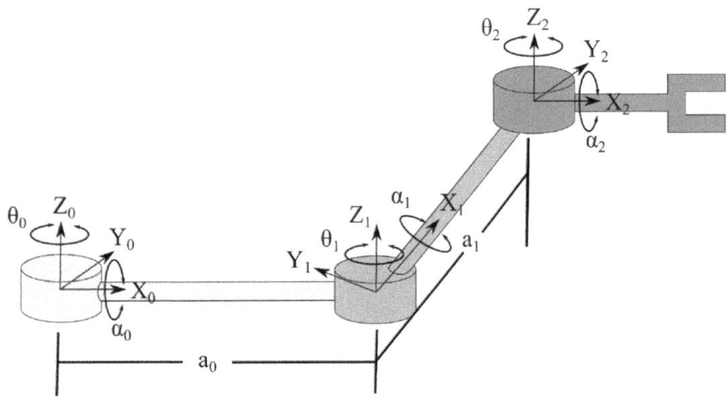

FIG. 2.9 A planar 3R mechanism with the end effector mounted to the final joint.

TABLE 2.6 The Denavit-Hartenberg parameter table for the planar 3R mechanism.

i	α_{i-1}	a_{i-1}	θ_i	d_i
0	0	0	θ_0	0
1	0	a_0	θ_1	0
2	0	a_1	θ_2	0

for this mechanism is shown in Table 2.6. Note that the α and d values are all at zero because this robot is confined to planar movements.

Using the values from the DH table, the homogenous transforms for each joint are calculated using the respective matrices from Table 2.3, as shown in Eqs. (2.36)–(2.38).

Joint 0 homogeneous transform matrix

$$Z(\theta_0, 0) = \begin{bmatrix} \cos\theta_0 & -\sin\theta_0 & 0 & 0 \\ \sin\theta_0 & \cos\theta_0 & 0 & 0 \\ 0 & 0 & 1 & 0 \\ 0 & 0 & 0 & 1 \end{bmatrix} \tag{2.36}$$

Joint 1 homogenous transform matrix

$$Z(\theta_0, 0)X(0, a_0)Z(\theta_1, 0) = \begin{bmatrix} \cos(\theta_0 + \theta_1) & -\sin(\theta_0 + \theta_1) & 0 & a_0\cos\theta_0 \\ \sin(\theta_0 + \theta_1) & \cos(\theta_0 + \theta_1) & 0 & a_0\sin\theta_0 \\ 0 & 0 & 1 & 0 \\ 0 & 0 & 0 & 1 \end{bmatrix} \tag{2.37}$$

Joint 2 homogeneous transform matrix

$$Z(\theta_0 0)X(0\mathrm{a}_0)Z(\theta_1 0)X(0\mathrm{a}_1)Z(\theta_2 0)$$
$$= \begin{bmatrix} \cos(\theta_0+\theta_1+\theta_2) & -\sin(\theta_0+\theta_1+\theta_2) & 0 & \mathrm{a}_0\cos\theta_0+\mathrm{a}_1\cos(\theta_0+\theta_1) \\ \sin(\theta_0+\theta_1+\theta_2) & \cos(\theta_0+\theta_1+\theta_2) & 0 & \mathrm{a}_0\sin\theta_0+\mathrm{a}_1\sin(\theta_0+\theta_1) \\ 0 & 0 & 1 & 0 \\ 0 & 0 & 0 & 1 \end{bmatrix} \tag{2.38}$$

Since the tool lies directly on Joint 2, its coordinates are the $\vec{d}$ displacement vector of the Joint 2 homogenous transform matrix, which has been extracted in Eq. (2.39).

$$\vec{d} = \begin{bmatrix} x \\ y \\ z \end{bmatrix} = \begin{bmatrix} \mathrm{a}_0\cos\theta_0+\mathrm{a}_1\cos(\theta_0+\theta_1) \\ \mathrm{a}_0\sin\theta_0+\mathrm{a}_1\sin(\theta_0+\theta_1) \\ 0 \end{bmatrix} \tag{2.39}$$

The translation Jacobian can then be calculated by substituting the values from $\vec{d}$ into Eq. (2.16), as shown in Eq. (2.40).

$$J_v = \begin{bmatrix} \frac{\partial x}{\partial\theta_0} & \frac{\partial x}{\partial\theta_1} & \frac{\partial x}{\partial\theta_2} \\ \frac{\partial y}{\partial\theta_0} & \frac{\partial y}{\partial\theta_1} & \frac{\partial y}{\partial\theta_2} \\ \frac{\partial z}{\partial\theta_0} & \frac{\partial z}{\partial\theta_1} & \frac{\partial z}{\partial\theta_2} \end{bmatrix} = \begin{bmatrix} -\mathrm{a}_0\sin\theta_0-\mathrm{a}_1\sin(\theta_0+\theta_1) & -\mathrm{a}_1\sin(\theta_0+\theta_1) & 0 \\ \mathrm{a}_0\cos\theta_0+\mathrm{a}_1\cos(\theta_0+\theta_1) & \mathrm{a}_1\cos(\theta_0+\theta_1) & 0 \\ 0 & 0 & 0 \end{bmatrix} \tag{2.40}$$

The R_z for each joint is $\langle 0,0,1\rangle$ because there is no translation along the Z axis, so the term is equivalent to the basis vector for the world Z axis. The rotational Jacobian, therefore, is calculated using Eq. (2.34), as shown Eq. (2.41).

$$J_\omega = [\zeta_1 R_{z1} \;\; \zeta_2 R_{z2} \;\; \zeta_3 R_{z3}] = \begin{bmatrix} 1\begin{bmatrix} 0 \\ 0 \\ 1 \end{bmatrix} & 1\begin{bmatrix} 0 \\ 0 \\ 1 \end{bmatrix} & 1\begin{bmatrix} 0 \\ 0 \\ 1 \end{bmatrix} \end{bmatrix} = \begin{bmatrix} 0 & 0 & 0 \\ 0 & 0 & 0 \\ 1 & 1 & 1 \end{bmatrix} \tag{2.41}$$

The full Jacobian is shown in Eq. (2.42).

$$J = \begin{bmatrix} -\mathrm{a}_0\sin\theta_0-\mathrm{a}_1\sin(\theta_0+\theta_1) & -\mathrm{a}_1\sin(\theta_0+\theta_1) & 0 \\ \mathrm{a}_0\cos\theta_0+\mathrm{a}_1\cos(\theta_0+\theta_1) & \mathrm{a}_1\cos(\theta_0+\theta_1) & 0 \\ 0 & 0 & 0 \\ 0 & 0 & 0 \\ 0 & 0 & 0 \\ 1 & 1 & 1 \end{bmatrix} \tag{2.42}$$

Substituting J into Eq. (2.35) yields the full equation relating the Cartesian space velocities to the joint space velocities, as shown in Eq. (2.43).

$$\begin{bmatrix} \dot{x} \\ \dot{y} \\ \dot{z} \\ \dot{\phi}_x \\ \dot{\phi}_y \\ \dot{\phi}_z \end{bmatrix} = \begin{bmatrix} -\mathrm{a}_0 \sin\theta_0 - \mathrm{a}_1 \sin(\theta_0+\theta_1) & -\mathrm{a}_1 \sin(\theta_0+\theta_1) & 0 \\ \mathrm{a}_0 \cos\theta_0 + \mathrm{a}_1 \cos(\theta_0+\theta_1) & \mathrm{a}_1 \cos(\theta_0+\theta_1) & 0 \\ 0 & 0 & 0 \\ 0 & 0 & 0 \\ 0 & 0 & 0 \\ 1 & 1 & 1 \end{bmatrix} \begin{bmatrix} \dot{\theta}_1 \\ \dot{\theta}_2 \\ \dot{\theta}_3 \end{bmatrix} \tag{2.43}$$

In the equation, x, y, and z represent translations along the X, Y and Z axes, respectively, while ϕ_x, ϕ_y, and ϕ_z similarly represent rotations. Additionally, θ represents the joint velocity. An easy double check shows that the robotic arm cannot move out of the plane given that $\dot{z} = 0$, $\dot{\phi}_x = 0$, and $\dot{\phi}_y = 0$. It is also known that $\phi_z = \theta_1 + \theta_2 + \theta_3$ and thus $\dot{\phi}_z = \dot{\theta}_1 + \dot{\theta}_2 + \dot{\theta}_3$.

Singularities

There exist joint locations where the robot Jacobian is not invertible, and these locations are called singularities. This occurs when there is alignment between two joints of the robot, as shown in Fig. 2.10. As the robot approaches singularity positions, the ratios for input velocities to output velocities approach infinity. This essentially means that movement is blocked while in that position. If a robotic system has singularities, typically the planned paths will avoid passing through the singularities. Another option is to drive the joints directly without checking the end effector velocities. However, this does not work for AM systems because it is critical to control the velocity of the end effector during printing.

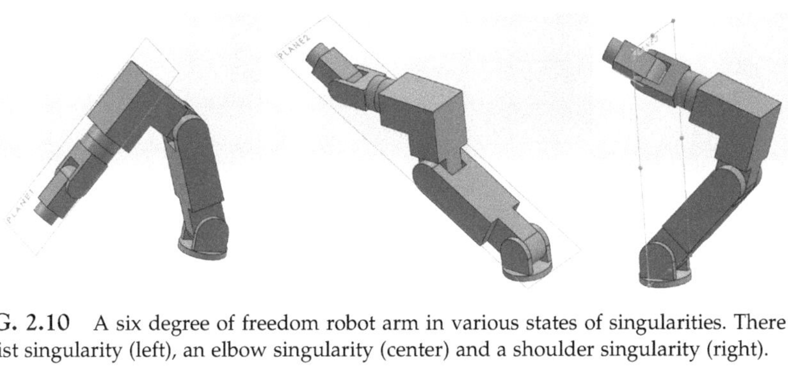

FIG. 2.10 A six degree of freedom robot arm in various states of singularities. There is wrist singularity (left), an elbow singularity (center) and a shoulder singularity (right).

Conclusion

Kinematics is the study of motion and is fundamental to the field of robotics and, in turn, additive manufacturing. This chapter has introduced the reader to the terminology and methods used when performing kinematic analyses of robotic systems. It has attempted to provide a basic understanding of a robot's kinematics, workspace, and Jacobian. These concepts are critical in the design and use of robotic platforms.

Robot kinematics can be mathematically described using rotation matrices or quaternions to define the transformations of each successive frame of a mechanism. Forward kinematics start with the ground link position and move through successive frames until the end effector is reached. If the dimensions of each link, and the displacement of each joint are known, the end effector position and orientation can be easily calculated. The collection of all possible end effector positions, determined by the forward kinematics, is the robot's workspace. Inverse kinematics start with the location of the end effector and work backwards through the preceding frames to determine which transformations place the end effector at the specified location. These are used to plan the robot's movement through its workspace and help ensure no positional singularities are encountered during the robot's motion. A robot Jacobian defines the relationship between the robot's kinematics, speed, and resulting forces.

With this foundation, the reader should be able to understand the basics of robot kinematics and be able to dive deeper into specific topics as needed when designing and path planning for their own additive manufacturing platform.

References

[1] J.M. McCarthy, G.S. Soh, Geometric Design of Linkages, vol. 11, Springer Science & Business Media, 2010.

[2] L.-W. Tsai, Robot Analysis: The Mechanics of Serial and Parallel Manipulators, John Wiley & Sons, 1999.

[3] J. Denavit, R.S. Hartenberg, A kinematic notation for lower-pair mechanisms based on matrices, J. Appl. Mech. 22 (2) (1955) 215–221, https://doi.org/10.1115/1.4011045.

CHAPTER

3

Motion platforms and kinematic arrangements

Peter Wang[a], *Abby Barnes*[a], *Alex Roschli*[a], *Eric MacDonald*[a,b], *Thomas Feldhausen*[a], *and Michael Borish*[a]

[a]Manufacturing Demonstration Facility, Oak Ridge National Laboratory, Knoxville, TN, United States, [b]University of Texas at El Paso, El Paso, TX, United States

Abstract

Within a machine, mechanisms and motion are organized in what is known as a "kinematic arrangement," which helps classify machines based on how they move. The most common kinematic arrangements for additive manufacturing systems are Cartesian, followed by delta, and then six-degrees-of-freedom robotic arms. However, there are a multitude of less common systems, such as the SCARA, polar robots, cable driven parallel robots, mobile platforms, and multi-agent systems. This chapter surveys these various kinematic arrangements to give a broad understanding of the mechanisms underlying motion within additive manufacturing systems. Understanding these mechanisms and their resulting motion provides a framework for discussing path planning for all scales and families of additive manufacturing.

Introduction

The collection of components that produce a machine's motion are often called the machine's "motion platform." Motion platforms have actuators, which are the components that are directly connected to and provide motion for the end effector of the machine. In the context of additive manufacturing (AM), actuators enable the end effector—usually a printhead or energy source—to reach the appropriate location at the appropriate time for deposition. Motion platforms can be described in terms of the organization of their components, termed a "kinematic arrangement" or

https://doi.org/10.1016/B978-0-443-15286-3.00001-1

a "kinematic configuration." There are a wide variety of kinematic arrangements, each of which have different advantages and disadvantages for manufacturing an object according to its specifications. These specifications may introduce requirements for build volume, AM method, material properties, print head size, end effector precision or other end effector forces, all of which affect the most suitable choice of kinematic arrangement.

This chapter surveys the kinematic arrangements that are seen in AM systems on the market right now, as well as some configurations that only exist in research environments. The goal of this chapter is to provide a general familiarity with the types of kinematic configurations that are available, their resulting motion, and the types of objects that each configuration is best suited to manufacture. The kinematic arrangements discussed in this chapter are Cartesian, delta, polar, SCARA, six-degrees-of-freedom robot arm, parallel cable driven robots, mobile platforms, and multi-agent platforms. Note that this discussion is not meant to provide an exhaustive list of every AM machine that exists, but rather establish a framework for AM kinematics that can be used to aid discussions, decisions, and development within the context of AM, including (but not limited to) path planning.

Cartesian systems

Cartesian systems are characterized by prismatic joints and linear, orthogonal actuators that act independently from each other, so when one actuator is moved, only one of the X, Y, or Z coordinate values will change. This makes the forward kinematics of the system very simple; each X, Y, and Z coordinate value is equal to the displacement of one actuator, plus an offset. The value of the offset depends on where the origin of the current coordinate frame is located within the machine's build volume. For example, if the origin is at the zero position of each actuator, then the coordinate values will be equal to the actuator displacement values. However, if the origin is at the center of the bottom (zero of the Z actuator) of an L x L build volume, the offset will be $-\frac{L}{2}$ and the coordinate values will be $\left(X_{actuator} - \frac{L}{2}, Y_{actuator} - \frac{L}{2}, Z\right)$. This simplifies the controls and minimizes the computational overhead from kinematic calculations. The simplicity of Cartesian systems has made this kinematic configuration the most common platform in AM and the most plentiful on the market.

Cartesian systems can be found in all families and scales of AM, ranging from systems that print tiny electrical circuits to systems that print concrete houses. They can be categorized as one degree-of-freedom (DOF), three DOF, or infinite build, based on which type of mechanisms are used as the actuators. Each category has a specific use in AM, as detailed in the following subsections.

One degree of freedom

The motion of a single degree-of-freedom Cartesian system is limited to translations along the Z axis. With this configuration, control over the X and Y axes typically comes from an energy source that scans a reservoir of loose material to melt or bind the material in specific locations. The energy source is typically mounted in a fixed position; mirrors are used to direct the energy to an appropriate place within the reservoir.

This kinematic arrangement is seen in vat polymerization AM systems, such as stereolithography (SLA), digital light processing (DLP) or liquid crystal displace (LCD) 3D printing. These three processes use a vat of resin that hardens when exposed to light of a specific intensity and wavelength. After each layer is hardened, the build plate moves one layer thickness away from the energy source and the process is repeated such that the newly hardened layer will bond to the previous layer. The energy source is usually mounted below the resin vat and the actuator moves upwards during the build process. The difference between the processes is how the light exposure is controlled. SLA uses mirrors to move a laser point along the appropriate paths and locations on the surface. The DLP system uses a laser projector to project an image of the entire layer at a single time. The LCD systems use an array of UV lights with an LCD screen above them to mask the negative space of the layer. All of these systems have a very similar physical appearance. The Formlabs Form 3 SLA printer [1] and the AnyCubic Photon Ultra DLP printer [2] are examples of resin-based printers with a one degree-of-freedom motion platform that moves upwards as the build progresses.

The single degree-of-freedom kinematic arrangement is also seen in powder bed fusion AM systems, such as selective laser sintering (SLS) and selective laser melting (SLM). In these systems, the powder bed (and the object being built) moves downwards and powder is added on top, burying the part as shown in Fig. 3.1. The energy source—a laser or an electron beam—is mounted above the platform and guided by using moving mirrors or electric fields, respectively. The primary difference between the two processes is that SLS sinters the powder, whereas SLM fully melts the powder [3]. The powders can be either metal or plastic.

In general, the direction of motion for powder bed fusion systems is opposite of that for the resin-based systems: powder beds move down, while resin vats move up. However, some large-scale resin-based systems had adopted the descending bed concept, such as the Mammoth from Materialize. This system implements a descending bed system in which

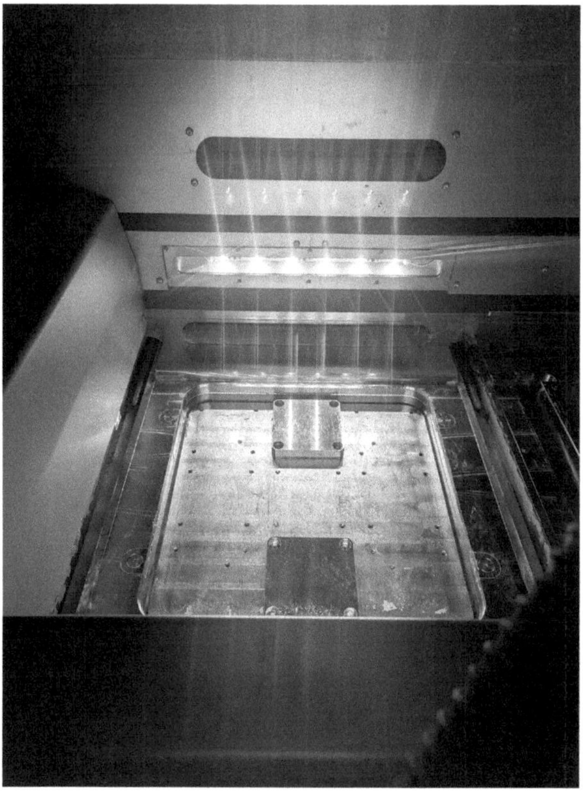

FIG. 3.1 The Add-Up system uses a descending powder bed, shown here without any powder and two build plates mounted to the bed.

the part descends into a large vat of resin as the surface is scanned by a laser [4]. This reversion is helps large-scale systems deal with the weight of the part; its much more difficult to hold a heavy part up high than low to the floor. With a descending vat, the part gets closer to the floor as it gets heavier, instead of getting further away.

Three degrees of freedom

Three degree-of-freedom (3DOF) Cartesian systems have actuators moving along all three X, Y and Z axes. Each actuator is typically linear, so the motion in each direction is decoupled such that moving one actuator only affects one axis of motion. This simplifies both the construction and controls of these types of systems. The scales of 3DOF Cartesian systems range from printing electronic circuit boards to those that can print houses. There is a COBOD concrete material extrusion printer that has

FIG. 3.2 The COBOD system shown here is used to print wind turbine towers using a concrete mix. *From Kamp C: Cobod International. COBOD, 10 June 2022, https://cobod.com/projects-partners/kamp-c/.*

printed a house, in which the house remained stationary and the printhead moved in 3DOF [5]. COBOD has used their system to print wind turbine towers in partnership with GE (Fig. 3.2). Additionally, attachments can be added to the end effector to add additional degrees of freedom or other capabilities, depending on the needs of the manufacturing process. The ease of control and construction, combined with the flexibility in scale and configuration, has driven the wide-scale adoption of this kinematic configuration. This configuration can be found widely in material extrusion, material jetting, binder jetting, and directed energy deposition, with a few seen even in sheet lamination and powder bed fusion systems.

The 3DOF systems can be further categorized into four different arrangements, based on the ratio of the degrees of freedom given to the printhead vs the base: one of them has either zero, one, two, or three degrees of freedom, while the other has the remaining degrees of freedom. The degrees of freedom are often divided between the two based on the size of the printhead and the size of the part. For example, a printer with zero degrees of freedom on the printhead (i.e., the platform has all three degrees of freedom) is advantageous for printing a very small part with a massive printhead. However, to print a house, the opposite is true: it's more efficient for the printhead to have all three degrees of freedom while the platform has none.

Isolating motion to one component with a three-to-zero ratio is not the only possible configuration. Motion along the three Cartesian directions (X, Y, and Z) can be assigned to the print head and the build platform in any ratio. One possible ratio gives two degrees of freedom (2DOF) to the printhead and a single degree of freedom (1DOF) to the platform. There are three possible configurations using this ratio. In terms of "(printhead, platform)," they are (XY, Z), (XZ, Y), and (YZ, X). However, upon examination it can be seen that the (XZ, Y) and (YZ, X) configurations lead to identical machines simply rotated 90 degrees about the Z axis, so these options collapse into a single configuration. The remaining (XY, Z) and (XZ, Y) configurations differ in that (XY, Z) builds objects upwards, using horizontal layers, while (XZ, Y) builds objects outwards, using vertical layers.

Binder jet is one AM process that uses the (XY, Z) configuration. It would be challenging to rapidly move the massive powder bed of a binder jet system in the X and X directions, because of both the large mass and the potential disturbance to the powder, so XY movement is done by the printhead. Additionally, the descending movement of the bed in the Z direction eliminates the need to build a powder containment structure around the part, which would be necessary if the bed was stationary and the printhead was moving. An example of an ExOne binder jet system that has an (XY, Z) is shown in Fig. 3.3 [6].

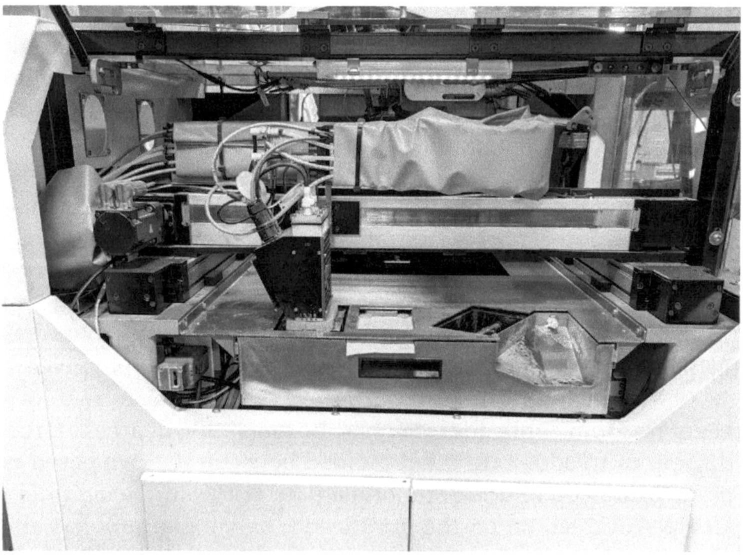

FIG. 3.3 An ExOne binderjet printer. The print head here moves in the X and Y directions while the powder bed lowers one layer at a time in the Z direction. *From "ExOne 2021: Our Catalog of Industrial 3D Printers and Comprehensive AM Services." ExOne, https://www.exone.com/en-US/ExOne-2021-Corporate-Brochure.*

The (XZ, Y) configuration is much less common, because the vertical layers require the printed material to withstand the force of gravity immediately after it has been printed. However, if this can be achieved, the (XZ, Y) configuration allows long and heavy parts to be printed without the need for an extremely tall printer. This configuration also allows objects with overhangs to be handled differently, potentially eliminating the need for support structures. A Thermwood printer has demonstrated this configuration, building vertical layers as the printhead moves in the X and Z directions while the part is horizontally moved away from the printhead in the Y direction [7].

The other possible two-to-one arrangement gives 1DOF to the printhead and 2DOF to the build platform. Again, the three cartesian directions, X, Y, and Z, can be assigned to the printhead and the build platform in any combination. The three possible combinations for movement, again in terms of (printhead, build platform), are (X, YZ), (Y, XZ) and (Z, XY). Once again, the (X, YZ) and (Y, XZ) configurations are essentially identical machines, so of those two, only the (X, YZ) configuration will be considered.

Examples of the (Z, XY) configuration are seen often when the printhead is massive or complex and the printed parts are relatively small. The Xerox ElemX system, shown in Fig. 3.4, uses this configuration [8]. The (X, YZ) configuration has not been seen. This configuration would involve a part mounted to a platform that is vertically cantilevered, requiring newly printed layers to resist the force of gravity, which no currently used material can do. Most prints that require a horizontal build can be printed with an (XZ, Y) configuration.

Infinite builds

An infinite build configuration is a variation of the other Cartesian arrangements in which a conveyor belt is used as one of the actuators. Conceptually, one dimension of the part is continuously and infinitely moved away from the printhead. The printhead is usually mounted at an angle, so that the layers will be angled relative to the conveyor belt. The easiest arrangement to conceptualize is the (XZ, Y) configuration with a conveyor belt as the Y actuator. With that configuration, the length (i.e., the Y dimension) of the part can be as large as afforded by the space and support beyond the conveyer belt. A prototype of this configuration was demonstrated by the Stratasys H2000 in 2017. The printhead on the Stratasys H2000 moves in the X and Z directions while the part moves away along the Y axis, and the back wall moves backwards with the part [9].

The printhead can also be angled to improve support of overhangs. When the printhead is angled, motion along the Y and Z axes is coupled.

FIG. 3.4 The Xerox ElemX system where the print head is very large and moves only in the Z direction. The parts are much smaller and move in the X and Y directions. *From "One Drop at a Time: Xerox 3D Prints With Liquid Metal." Engineering.com, https://www.engineering.com/story/one-drop-at-a-time-xerox-3d-prints-with-liquid-metal.*

This means that moving the printhead in either the Y or Z directions results in a simultaneous change in both its Y and the Z coordinates. Motion along the X axis is still independent. An example of an infinite (XZ,Y) configuration with an angled printhead is shown in Fig. 3.5. This configuration has also been demonstrated on by the Blackbelt 3D Printer, which has printed objects 2 m long [10]. Additionally, this configuration can be used for continuous production of smaller parts, eliminating the need to remove parts after every build.

FIG. 3.5 The Powerbelt 3D shown here prints on a 45-degree plan with a conveyor belt which gives it a theoretically infinite build volume in the direction of the conveyor belt (Y-direction). *From K. Stevenson, "The blackbelt 3D printer story." Fabbaloo, 23 July 2020, https://www.fabbaloo.com/2018/12/the-blackbelt-3d-printer-story.*

Delta configurations

Delta robots have three or four sets of parallel mechanisms, called linkages, that drive the end effector with three or four degrees of freedom, respectively. The actuators for a delta robot can be either rotating or linear. An example of a delta robot with parallel linear actuators is shown in Fig. 3.6. Delta robots are able to move very quickly and precisely because of the fact that all axes of motion are linked by the linkages. In the manufacturing industry, delta robots are typically used for fast pick-and-place operations. In AM, a three-linkage configuration with linear actuators is used in systems such as the BOSSdelta 300 from SeeMeCNC [11].

The primary advantage of delta robots is that the linkages are very light weight because the drive motors can be mounted at the base of the mechanism. However, this generally also causes delta robots to have a limit on the weight of the end effector. Additionally, delta robots have a workspace that is generally limited to the area directly beneath the end effector, similar to a Cartesian configuration. However, unlike a Cartesian robot, there will be a significant amount of unused space above the upper most position of the end effector.

This configuration can be scaled to very large sizes. For example, the Wasp BigDelta printer is 12 m tall, 7 m wide, prints clay and straw, and has been used to print a 40-ton structure [12]. The delta configuration

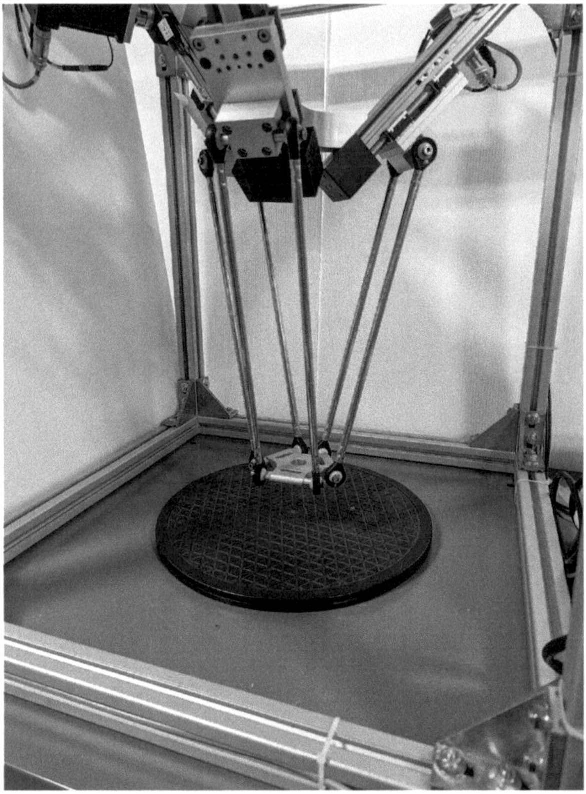

FIG. 3.6 The delta robot shown here is made by Igus and is actuated by three linear actuators and has three degrees of freedom. *From "All about Bigdelta Wasp." 3D Printers WASP, 30 Jan. 2017, https://www.3dwasp.com/en/giant-3d-printer-bigdelta-wasp-12mt/.*

may scale more easily than the Cartesian configuration in the vertical direction. This is because the end effector and the linkages will be a lower mass than the X and Y gantries of a Cartesian robot, and the linkages will also be much stiffer than the Cartesian robot's single Z pole.

Six DOF robot arm configurations

Six degree-of-freedom (6DOF) robotic arms are serial chain robots that typically have six revolute joints, ultimately giving the end effector six degrees of freedom. Some AM systems use a 6DOF robot arm as the only actuator for the end effector. For example, the Ex-One S-Max Flex sand binderjet printer is one AM system in which the print head is mounted on a 6DOF robotic arm [13]. The Branch Technology polymer extruder

FIG. 3.7 Left: The Lincoln Electric wire arc additive manufacturing system, which has a robotic build plate with two degrees of freedom. Right: The GKN laser wire system, which has its deposition system mounted on a 6DOF robot arm and a 2DOF positioner for the build plate. *From "Lincoln Electric and SES Combine Additive Manufacturing Expertise to Grow Market." Lincoln Electric, 2 Nov. 2021, https://www.lincolnelectric.com/en/Newsroom/Press-Releases/2021/11/Lincoln-Electric-and-SES-Combine-Additive-Manufacturing-Expertise-to-Grow-Market.*

is also mounted on a 6DOF robot arm and has been successfully used to print very complex shapes [14].

Other systems use a 6DOF arm with a rotating platform or a trunnion for the build plate, allowing the build plate to be rotated with one or two degrees of freedom. An example of this is the Lincoln Electric AM platform, which has an MIG welding torch mounted on a 6DOF arm and uses a two degree-of-freedom robot for the base plate, as shown on the left side of Fig. 3.7 [15]. Another example is the GKN laser wire deposition system, which has the ended effector mounted on a 6DOF arm and uses a one degree-of-freedom trunnion for the base plate, as shown on the right side of Fig. 3.7 [16]. Other systems mount the 6DOF robot on a track, allowing it to translate and significantly increasing the size of the workspace.

6DOF robot arms are commonly seen in assembly lines welding, painting, inspecting, and performing pick and place operations. Within AM, robotic arms have been used in laser wire, wire arc, binder jet, and polymer processes. The degrees of freedom of this configuration enable off-axis printing because the orientation of the end effector can be rotated about any axis within the build space. This can also enable the direct printing of complex 3D structures, such as those created by Branch Technologies [14]. An additional benefit of using 6DOF robotic arms is that the controls are well developed, and the robotic arms are widely available, which drives both purchase and development costs down when compared to using a custom-designed motion platform.

Configurations in research and development

The kinematic arrangements discussed so far are well-established and commercialized. This section discusses configurations that are still under development and are not as commonly seen in AM. These are examples of creative ways that the kinematics of AM systems can be configured.

Polar configurations

The polar or cylindrical configuration has the same configuration as a Cartesian robot, but with one prismatic joint replaced by a revolute joint. In an additive platform, a revolute joint is often added to the Z axis actuator, allowing the print bed and/or printhead to rotate about and translate along the Z axis. The prismatic joint that this revolute joint replaces is from either the X or Y axis actuator. The three resulting movements (one rotational and two translational) give the full three degrees of freedom necessary to manufacture components and are defined in a cylindrical coordinate frame. The remaining prismatic joint allows translation along the radius of the cylinder. The rotation enables printing in continuous spiral or helical paths, reducing or eliminating starts and stops. This configuration is ideally suited for printing cylindrical or conical components but can also be used to produce many smaller components on the build table.

One example of this configuration is the continuously rotating powder bed fusion system developed by Carter et al., which utilizes a laser system that moves vertically and radially, as shown in Fig. 3.8 [17]. On a much larger scale—big enough to build houses—is the Apis Cor concrete printer, which has an arm that rotates about the base and moves up and down, as well as a print head that moves radially along the arm [18]. Another example is the DP Polar machine, in which the entire print bed rotates and moves downwards while the printhead spans the diameter of the print area and remains stationary [19]. The DP Polar machine is targeting commercial production of many small components but has not been widely adopted.

SCARA configurations

The Selective Compliance Assembly Robot Arm (SCARA) is a serial chain robot that is traditionally characterized by two sequential revolute joints, followed by one prismatic joint. Generally, the two revolute joints rotate around the same axis, working together to provide translation along the other two axes, while the prismatic joint moves up and down the axis of rotation. Together, the joints provide three degrees of freedom. A third revolute joint can be added after the prismatic joint for a fourth degree of freedom, as seen in an ABB SCARA [20]. Like a polar configuration, the workspace of a SCARA is a cylinder with an axial hole. The X and Y Cartesian coordinates of the end effector are linked such that the

FIG. 3.8 The figure above shows the ORNL rotary laser powder bed fusion system with two lasers that move vertically upwards and a rotating build platform. *From W.T. Carter, et al., "A direct metal laser melting system using a continuously rotating powder bed." 2019 International Solid Freeform Fabrication Symposium. University of Texas at Austin, 2019.*

movement of either revolute joint will drive a change in both the X and Y coordinates. Like 6DOF robot arms, SCARA are used in factory settings for fast pick and place operations.

In AM, modifications have been made to the traditional revolute-revolute-prismatic joint configuration of SCARA systems. For example, the Sculpto printer uses a SCARA with a revolute-prismatic-revolute configuration, placing the prismatic joint between the two revolute joints. The build plate rotates while the prismatic and the other revolute joint move the polymer filament in an arc [21]. From the frame of the build plate, the kinematics look the same as a traditional SCARA system. Others have reversed the traditional configuration, using only one prismatic joint and two revolute joints, calling it the Dual-Disc Polar Printer [22]. The prismatic joint holds the printhead, while the two revolute joints hold the build table.

Cable-driven platforms

Cable Driven Parallel Robots (CDPRs) are parallel platforms whose end effectors are positioned using a series of cables. In these systems, the actuators have been replaced by cable winders and thus are unable to push the end effector into position, instead relying on cable tension to pull and hold the end effector. The control of the tension on the cables can be complicated, and these systems are commonly over-constrained. There must be tight coordination between the cable winders to prevent back-driving any actuators. Additionally, understanding the modal vibration frequencies of the systems is critical to prevent excessive vibration due to any movements of the end effector or base actuators.

The advantage of CDPRs is that they enable a very large workspace. Such systems have been used to control massive structures, such as a 500-m telescope in China [23] and the SkyCam [24] in sports arenas, both of which have workspaces that cover hundreds or thousands of square meters. This scalability can enable the printing of extremely large structures, and in AM, CDPRs have been used in polymer and concrete extrusion systems. For example, the Hangprinter has printed a 4 m tall tower from polymer filament [25]. Additionally, the SkyBAAM printer has printed a concrete spiral staircase that weighed several thousand pounds, shown in Fig. 3.9 [26]. CDPR AM platforms have been studied by makers, academia, and research institutions with a multitude of different cable configurations but have not been commercialized to date.

FIG. 3.9 ORNL's SkyBAAM is shown printing a concrete spiral staircase outside with the winding stations contained inside a shipping on container, with vertical support coming from a forklift. *From P.C. Chesser, et al., Kinematics of a cable-driven robotic platform for large-scale additive manufacturing, J. Mech. Robot. 14 (2) (2022).*

Mobile platforms

Mobile printing platforms carry a printhead on top of a vehicle or other mobile platform. They can be ground-based, using a crawler or wheeled vehicle, or they can be airborne, using a drone. The printhead is often connected to the mobile platform using a robotic actuator. The Cybe Robot Crawler, shown in the top portion of Fig. 3.10, is one example of a mobile platform that prints cement mortar using a 6DOF robotic arm attached to a crawling robot [27]. The In Situ Fabricator, shown in the bottom portion of Fig. 3.10, also uses a crawler and robotic arm set up for their "rock print" method, which uses alternating layers of coarse gravel and twine to construct structures [28]. An airborne platform was produced by the

FIG. 3.10 The Cybe Robotic Crawler (top) cement mortar printer, that has a 6DOF robotic arm mounted on a tracked crawling mobile base. The In Situ Fabricator system (bottom) uses a crawling printer for their "rock print" method, in which layers of gravel and twine are used to print structures. *From K. Dörfler, et al., Additive manufacturing using mobile robots: opportunities and challenges for building construction, Cem. Concr. Res. 158 (2022) 106772.*

University of Bath, using a quadcopter drone to carry a delta robot with a printhead that extruded mortar [29]. The fast movement of the delta robot enabled the end effector position to remain stable as the flying platform drifted.

The mobile platforms give these AM systems the ability to print objects larger than the printer itself, with a theoretically infinite build volume. The mobile platform also gives significant freedom for the geometries, as irregular or high-aspect-ratio shapes are no longer limited by the printer size. However, given this freedom of movement, an additional control system must be used to locate the mobile base in space and ensure accurate printing. A laser tracker or vision system with fiducial markers have both been established as options to track the position of the system. If the mobile platform is arial or working on uneven terrain, then the orientation of the system must also be considered and compensated for by the end effector. Ariel platforms have the additional complexity of high-sensitivity to environmental conditions, such as air flow and air temperature gradients, so the control system must be able to compensate for these uncertainties. Overall, mobile platforms are an attractive option for printing extremely large objects without the need for a correspondingly large robotic system, but the necessary control systems are still under development.

Multi-agent platforms

Multi-agent systems utilize multiple printheads to collaborate on a single build. The build can contain a single large part or multiple parts that are printed simultaneously. The coordination between the multiple printheads requires complex feedback control algorithms with sensors that enable each printer to be aware of the position of the other printers in the system. Zhang et al. has developed one such multi-agent system that uses robotic arms mounted on mobile platforms to print a cement mortar [30]. This system has been tested using two robotic agents, which are shown in Fig. 3.11.

Dividing the workload between the various printing agents optimizes throughput. With increasing research on flying AM systems and flying swarm cooperation algorithms, the doors are opening for massive, collaboratively constructed structures. The combination of the concepts of mobile platforms and multi-agent systems enables very high deposition rates and extremely large, printed components. These systems exhibit similar behaviors as those of insects such as ants, bees, and wasps. Scaling with this type of system is easy; if higher deposition rates are needed, then more printing agents can be added until the physical limitation on space is reached when the agents interfere with each other. However, maintaining

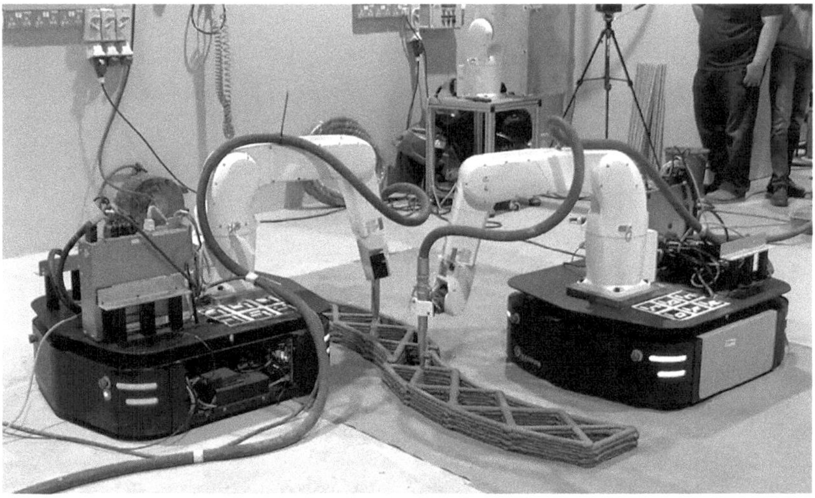

FIG. 3.11 Multiple mobile robotic printing systems collaborating to print a single cement mortar structure. *From X. Zhang, et al., Large-scale 3D printing by a team of mobile robots, Autom. Constr. 95 (2018) 98–106.*

print quality and consistent material properties are fields that still require significant research.

Conclusion

This chapter has surveyed the various kinematic arrangements that are seen in AM systems today and provides a necessary framework to examine the motion and path planning for an AM system. The kinematic arrangements of AM systems are being continuously developed and demonstrate that AM is still in its nascent stage. As AM technology matures, purposeful systems that are designed for specific applications and are kinematically unique will become increasingly more commonplace, following in the footsteps of subtractive manufacturing tools. The upcoming chapters of this book will discuss how to plan paths for these various systems, as well as some of the considerations for pathing based on the unique capabilities of each kinematic arrangement.

This chapter has also shown how kinematic arrangements provide an avenue to a future in which AM is not limited by print size, throughput or quality. Then, designers and manufacturers will be able to choose whether a part can and should be additively manufactured, instead of the limitations of AM choosing for them.

References

[1] "Form 3+: Industrial-Quality Desktop SLA 3D Printer." Formlabs, https://formlabs.com/3d-printers/form-3/.

[2] "Anycubic Photon Ultra." ANYCUBIC 3D Printing, https://www.anycubic.com/products/photon-ultra.

[3] S.R. Narasimharaju, et al., A comprehensive review on laser powder bed fusion of steels: processing, microstructure, defects and control methods, mechanical properties, current challenges and future trends, J. Manuf. Process. 75 (2022) 375–414. Stephanie. "6 Things You Didn't Know about 3D Printing a Woolly Mammoth.".

[4] I.materialise, 3D Printing Blog | I.materialise, 22 Jan. 2019. https://i.materialise.com/blog/en/6-things-you-didnt-know-about-3d-printing-a-woolly-mammoth/.

[5] Kamp C: Cobod International, COBOD, 10 June 2022. https://cobod.com/projects-partners/kamp-c/.

[6] "ExOne 2021: Our Catalog of Industrial 3D Printers and Comprehensive AM Services." ExOne, https://www.exone.com/en-US/ExOne-2021-Corporate-Brochure.

[7] G. Nehls, Thermwood demonstrates vertical layer print additive technology, 3 Sept. 2020, CompositesWorld. https://www.compositesworld.com/news/thermwood-demonstrates-vertical-layer-print-additive-technology.

[8] "One Drop at a Time: Xerox 3D Prints With Liquid Metal." Engineering.com, https://www.engineering.com/story/one-drop-at-a-time-xerox-3d-prints-with-liquid-metal.

[9] An in-Depth Look at Stratasys' New out-of-the-Box 3D Printing Tech. *Engineering.com*. https://www.engineering.com/story/an-in-depth-look-at-stratasys-new-out-of-the-box-3d-printing-tech.

[10] K. Stevenson, The Blackbelt 3D Printer Story, Fabbaloo, 23 July 2020. https://www.fabbaloo.com/2018/12/the-blackbelt-3d-printer-story.

[11] "Bossdelta™ 300 Desktop 3D Printer." SeeMeCNC, https://www.seemecnc.com/products/bossdelta-300-3d-printer.

[12] All about Bigdelta Wasp, 3D Printers WASP, 30 Jan. 2017. https://www.3dwasp.com/en/giant-3d-printer-bigdelta-wasp-12mt/.

[13] P. Hanaphy, et al., Rapid+TCT 2022: 3D Printing Innovators Break Down Manufacturing Barriers, 3D Printing Industry, 20 May 2022. https://3dprintingindustry.com/news/rapidtct-2022-3d-printing-innovators-break-down-manufacturing-barriers-209637/.

[14] Adlughmin., Branch Technology 3D Prints Building Walls With World's Largest Free-form 3D Printer—Launches 3D Printed Home Competition, 3DPrint.Com | The Voice of 3D Printing / Additive Manufacturing, 16 Oct. 2021. https://3dprint.com/85215/branch-3d-printed-walls/.

[15] Lincoln Electric and SES Combine Additive Manufacturing Expertise to Grow Market, Lincoln Electric, 2 Nov. 2021. https://www.lincolnelectric.com/en/Newsroom/Press-Releases/2021/11/Lincoln-Electric-and-SES-Combine-Additive-Manufacturing-Expertise-to-Grow-Market.

[16] GKN Opens World's Largest AM Pilot Production Cell at Ornl. *TechnicalTextile.net*. https://www.technicaltextile.net/news/gkn-opens-world-s-largest-am-pilot-production-cell-at-ornl-250079.html.

[17] W.T. Carter, et al., A direct metal laser melting system using a continuously rotating powder bed, in: 2019 International Solid Freeform Fabrication Symposium, University of Texas at Austin, 2019.

[18] Frank the Printer. "How Does Apis COR Differentiate Itself from Its Competitors?" Apis Cor Blog, https://blog.apis-cor.com/the-effects-of-cement-and-global-warming-0.

[19] V. Carlota, DP Polar Presents Its Ampolar I2 Solution for Industrial Scale Production, 3Dnatives, 29 Nov. 2019. https://www.3dnatives.com/en/dp-polar-ampolar-i2-solution-021220194/.

[20] ABB Expands Scara Robot Range for Faster, High Precision Assembly, News, 22 Sept. 2021. https://new.abb.com/news/detail/82302/prsrl-abb-expands-scara-robot-range-for-faster-high-precision-assembly.
[21] Sculpto Pro2, Sculpto, 12 Apr. 2021. https://sculpto.eu/.
[22] "Dual-Disc Polar Printer." Hackaday.io, https://hackaday.io/project/177633-dual-disc-polar-printer.
[23] N. Rendong, Five hundred meter aperture spherical radio telescope (FAST), Sci. China Ser. G Phys. Mech. Astron. 49 (2) (2006) 129–148.
[24] C. Scott, 3D Printing Outside the Box: The Hangprinter Prints in Midair Without Any Limits, 3DPrint.Com|The Voice of 3D Printing/Additive Manufacturing, 18 Oct. 2021. https://3dprint.com/167250/hangprinter-midair-3d-printer/.
[25] C. Chesher, Robots and the Moving Camera in Cinema, Television and Digital Media, International Workshop on Cultural Robotics, Springer, Cham, 2015.
[26] P.C. Chesser, et al., Kinematics of a cable-driven robotic platform for large-scale additive manufacturing, J. Mech. Robot. 14 (2022) 2.
[27] 3D Concrete Printers, CyBe Construction, 5 Oct. 2022. https://cybe.eu/3d-concrete-printing/printers/.
[28] K. Dörfler, et al., Additive manufacturing using mobile robots: opportunities and challenges for building construction, Cem. Concr. Res. 158 (2022) 106772.
[29] M. Clemens, Drones That Use 3D Printing for Building Construction and Repair, 3Dnatives, 23 Sept. 2022. https://www.3dnatives.com/en/drones-use-3d-printing-for-building-construction-and-repair-230920224/.
[30] X. Zhang, et al., Large-scale 3D printing by a team of mobile robots, Autom. Constr. 95 (2018) 98–106.

CHAPTER

4

Geometry data storage

Alex Roschli[a], Michael Borish[a], Abby Barnes[a], Eric MacDonald[a,b], Peter Wang[a], and Thomas Feldhausen[a]

[a]Manufacturing Demonstration Facility, Oak Ridge National Laboratory, Knoxville, TN, United States, [b]University of Texas at El Paso, El Paso, TX, United States

Abstract

3D-printing begins with the design of an object using computer aided design (CAD) software. The geometry of the object must be exported and saved in a data file format that can be used in the slicing process to generate machine instructions for printing. The standard method of saving the data is to tessellate the object as a triangulated mesh stored as a .stl file. Due to the flat triangular faces used to store the data, this mesh is a low-resolution representation of the high-fidelity object designed in CAD. The STL has its limitations with respect to geometrical accuracy, material information, and instances of invalid mesh data, but can still be used for the 3D-printing process. New file variants, such as OBJ, AMF, and 3MF, are being developed to fix some of these issues and increase the capabilities of geometry data storage for 3D-printing.

Introduction

Objects to be 3D printed must first be modeled with computer aided design (CAD) software. Many different software packages exist for generating 3D objects, such as Solidworks [1], FreeCAD [2], Autodesk Alias [3], and more. Each software has a unique approach to 3D modeling. For example, Solidworks uses solid body modeling, in which the object is created as a solid body and features can be added to or changed on the object. FreeCAD, on the other hand, utilizes parametric modeling, in which mathematical models and systems drive the shape and design of the object. Additional programs, such as Autodesk Alias, work through surface modeling, which allows the designer to shape and modify the surfaces of the object and later make it solid. Regardless of the software package and modeling approach,

https://doi.org/10.1016/B978-0-443-15286-3.00006-0

the end result must be the same: a solid body object which is used to represent all of the geometry data of the object to be printed.

Once a solid body has been generated in a CAD program, the geometry data must be transferred to the slicing process in a way that the slicer can handle. Slicing programs can't typically import the native CAD geometry data because, just as each modeling software has its own approach to 3D modeling, most CAD programs also have their own proprietary file format that stores their editable geometry data, such as the .sldprt file from Solidworks. This means that the geometry must be exported from the CAD program using a common (shared) file type that can be read by a slicing program.

The industry default file type is the STL file, with the .stl file extension. An STL file is a collection of vertices, edges, and faces that represents the surface geometry of a solid body, called a "mesh object," as shown in Fig. 4.1 [4]. STL isn't an acronym, but it is often said to stand for standard tessellation language. The tessellation, or tiling, of the solid body object's surface is completed with triangles during the STL generation process. Triangles are used to represent every surface of the solid body object. The sizing of the triangles is dependent on the resolution of the STL file. A high resolution STL will use smaller triangles, and thus more triangles, to represent the surface more accurately. A low resolution STL will use larger and fewer triangles, which represents the object less accurately. However, a high resolution STL is a larger file that takes longer to generate compared to a low resolution STL, which can be generated more quickly and creates a smaller file. In most CAD programs, during the STL creation and export, the user can define the STL resolution by adjusting the feature tolerance, which ultimately determines how many triangles are generated.

Although the STL is the industry standard, geometry limitations arise when geometry information is stored this way. With either a high resolution or low resolution STL, the stored geometry data is not a 100% accurate representation of the geometry modeled in the CAD software. This is because the triangles that are used to represent the surface are drawn with straight lines, which causes a loss of information on rounded or curved surfaces. Fig. 4.2 shows a perfect circle as would be drawn in CAD, as well

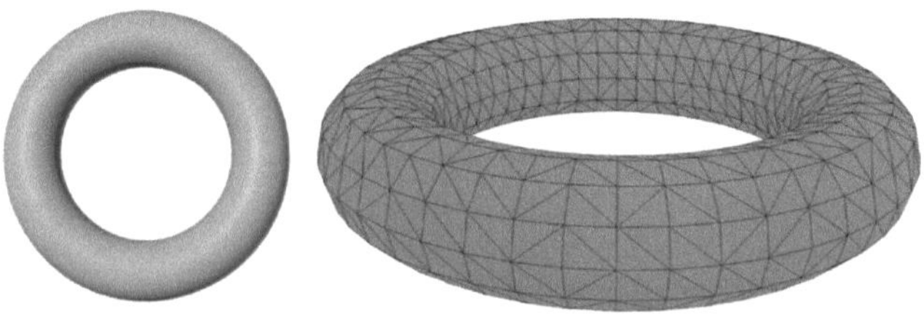

FIG. 4.1 A solid body object designed in CAD (left) and the resultant STL representation (right).

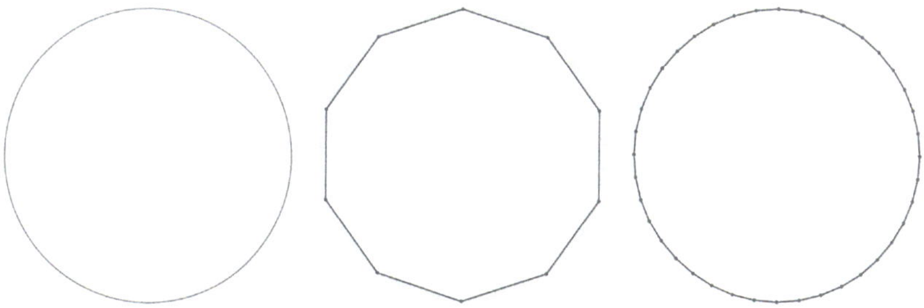

FIG. 4.2 A perfect circle (left), a circle approximated with 10 line segments (center), and a circle approximated with 40 line segments (right). The shaded areas indicate the amount of the original circle area left out of the STL representation.

as two representations (representing low and high resolutions) of that circle when drawn with straight lines.

It is clear that the represented area for each shape is not the same, with the straight-line representations covering less area than the perfect circle. The center object, with ten line segments, covers 6.45% less area than the perfect circle. This represents a low-resolution STL. The right object, with 40 line segments, covers 0.41% less area than the perfect circle, and represents a high-resolution STL.

To be 3D-printable, the resultant solid body STL object, as exported from the CAD program, must meet three requirements:

(1) The object must be manifold. A manifold object is a 3D object whose surface can be unfolded into a 2D object with all of its normals pointing in the same direction. Any normal that points in a different direction is called an inverted normal. Inverted normals make it difficult to determine the inside versus the outside of the object, which makes it difficult to define the volume of the object. A non-manifold object can be seen in Fig. 4.3.

(2) The object must have a measurable volume. Objects without a measurable volume, often surfaces that have zero thickness, cannot be

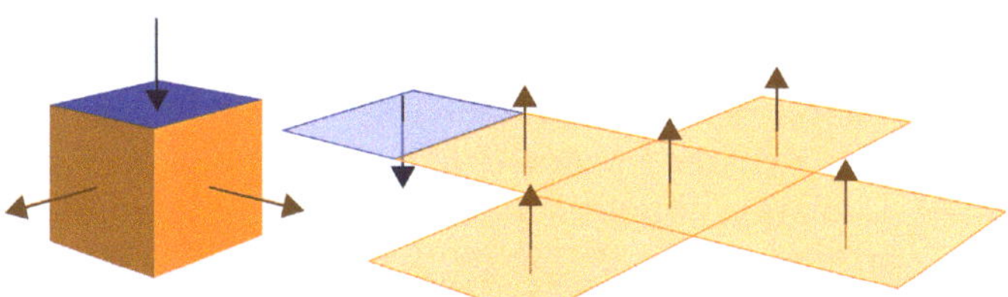

FIG. 4.3 A cube with an inverted normal, causing the face to point the wrong direction when unfolded.

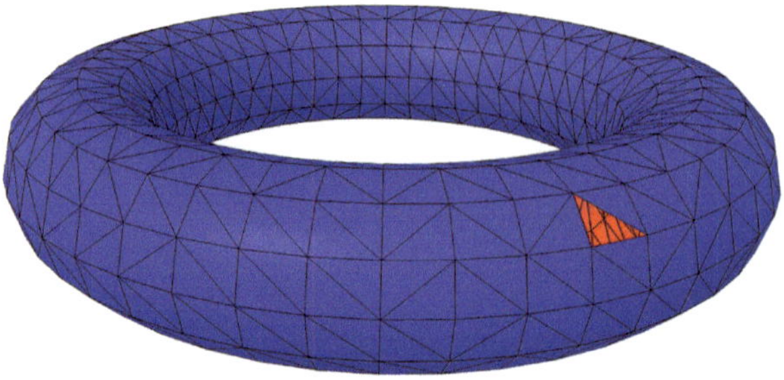

FIG. 4.4 A donut shape with a hole in its surface. The inside of the donut, which can be seen through the hole, is colored orange.

3D printed because there is no volume to represent with 3D-printed toolpaths.

(3) The object must be watertight. A watertight object is an object with no holes or pores in its surface. Holes in the surface prevent an object from being a solid. Instead, the object is seen as only a surface that is wrapped around empty space, without any measurable volume. Note that holes in the surface are different than holes in the object itself, which are acceptable. Fig. 4.4 illustrates the difference between surface holes and holes in objects.

If the resultant STL object does not meet all these requirements, additional software exists to repair the geometry by reversing normals that are inverted, thickening surface geometries that have zero thickness, and stitching holes that cause an object to not be watertight. The process of repairing an STL to make it 3D printable is called "healing" an STL.

STL file formats

STL files store data to represent every triangle used for the entire mesh. More specifically, the file contains the vertices for each triangle as well as the triangle's surface normal. The vertices and normals are represented in 3D space, with each having *X*, *Y*, and *Z* values (for vertices) or *I*, *J*, and *K* values (for normals). Only one mesh is captured within an STL, but the mesh can contain multiple bodies. STL files can be stored in ASCII (American Standard Code for Information Exchange) format or in binary format, which are discussed in more detail below. Both ASCII and binary do not store information about the units used in the file, making the file unitless. It is on the user to define the units when exporting the STL and use the same units when importing the STL file to another program, or to scale the imported STL file to achieve the desired size. The industry standard is to use millimeters as the unit.

ASCII

ASCII files follow a formatting standard using ASCII characters to represent text. An ASCII STL file uses this standard format to store the geometry data information in a way that can be read by a person. The downside of human readability is an increased number of characters, resulting in a larger file size compared to its binary counterpart. For example, a 6in. (152.4mm) cube creates an ASCII STL that is 3282 bytes, whereas the equivalent binary STL is just 684 bytes.

The format of an ASCII STL file begins with a first line that contains the string "solid" followed by the name of the mesh. If no name is given, the first line will simply contain "solid" followed by a space. The string "solid" is used as an opening tag to enclose the body. The body contains of all the geometry data defining the object's shape, and is stored as individual triangles, called facets.

Each facet is made up of seven lines of text, including the beginning and ending tags. The first line of a facet contains the string "facet normal" followed by the normal vector *I*, *J*, and *K* values, each separated by a space, and written as a floating-point number in exponential form. The second line contains just the string "outer loop." The third line represents the first vertex and includes the string "vertex" followed by the *X*, *Y*, and *Z* values of the vertex, each separated by a space, and written as a floating-point number in exponential form. The next two lines, lines four and five of the facet, follow the same format and are used to represent the second and third vertices, respectively. Following the third (and final) vertex line is the sixth line with the string "endloop," followed by the seventh, and last, line with the string "endfacet." This line concludes the definition of one facet. These seven lines are then repeated for every facet within the file. Below is an example of one facet from the 6in. (152.4mm) cube in ASCII format. The entire ASCII file can be seen in the appendix.

```
facet normal -1.000000e+00 0.000000e+00 0.000000e+00
    outer loop
        vertex 0.000000e+00 0.000000e+00 1.524000e+02
        vertex 0.000000e+00 1.524000e+02 1.524000e+02
        vertex 0.000000e+00 0.000000e+00 0.000000e+00
    endloop
endfacet
```

After the last facet, a single line with the string "endsolid" is used to denote the end of the file. Despite the use of the "solid" tag, the STL file isn't guaranteed to be solid. As previously mentioned, the STL needs to be solid for proper slicing for 3D-printing, but STLs aren't explicitly required to be solid during creation.

Binary

A binary file is not a text file; the data is stored within the file using a sequence of bytes. Because of this, the binary STL file, unlike the ASCII STL file, cannot easily be read by a person. The advantage of using the binary format is that the files are much smaller. The smaller size makes the binary format more common than the ASCII format, despite lacking human readability.

Binary STL files contain an 80-character header that can be used to name the mesh. This string shouldn't start with "solid" to avoid confusing a file importer that may assume it to be an ASCII STL file. After the header is a 4-byte unsigned integer representing the total number of facets, or triangles. The remainder of the file is used to define each triangle's vertices and normal vector, just like in the ASCII STL file. Each triangle uses 50 bytes of storage: 12 bytes for the normal, 12 bytes for each vertex, and 2 bytes for the attribute byte count, which is 0. When combined with the 80-character header, using 80 bytes, and the 4-byte number of triangles, the resultant binary STL file will always have a file size ending in 34 or 84 bytes. The file ends with the last triangle definition; there is nothing to signify the end of file. A full length binary STL file for a 6 in. (152.4 mm) cube can be seen in the appendix.

Other geometry storage file types

The STL file was first conceived in 1987 by the Albert Consulting Group. The STL was developed specifically for use in 3D-printing, alongside the development of the stereolithography process. At the time, computers were slow and storage was expensive. Therefore, the previously mentioned geometry limitations of the STL file were not a problem because more advanced and complex file types couldn't be properly handled by the computers anyway. However, as computer technology has advanced, the limitations of the STL file continue to become more and more of an issue. Since 1987, the format and functionality of the STL file has remained unchanged, which has led to development of new file variants. So far, three different file types have been released as alternative methods of storing geometry data: OBJ, AMF, and 3MF. One commonality of these new file types is that each vertex is indexed and stored only once, independently of faces. In general, this would result in a smaller file size and make these file types easier to parse. However, the impact on computations is probably negligible and the inclusion of other information, beyond faces and vertices, ultimately makes these files larger than an STL.

The OBJ file

The OBJ file, also known as the Wavefront OBJ file because it was developed by Wavefront Technologies around 1990, is often used for 3D graphics but can also be used in 3D-printing [5]. The OBJ file uses the file extension .obj and is capable of storing texture and color information because it was originally created for computer graphics, specifically video games and animations. Most 3D printers aren't able to make use of this data, but newer systems are starting to integrate support for multi-color printing.

Similar to the STL file, the OBJ file contains vertex data in three dimensions and doesn't store the units for the vertex coordinates. However, unlike in an STL, each of the vertices are stored in order, counterclockwise relative to the centroid when viewed from the exterior, which removes the need for the face normal. Additionally, the OBJ isn't limited exclusively to triangle representations of the mesh surface. OBJ files can use polygon faces with more than three vertices as well as free-form geometry with curves and NURBS surfaces. NURBS stands for non-uniform rational basis spline, which is a mathematical tool for representing splines in computer graphics. NURBS aren't currently used for creating geometries when 3D-printing. Most computer libraries used for 3D-printing from an OBJ file start by triangulating the OBJ file into an STL-like mesh.

OBJ files begin with a header to give the mesh a name and call out any associated files. The mesh name follows the "#" symbol, which denotes the beginning of a comment, i.e. code that doesn't execute. Associated files are used for material properties, and contain data stored within a .mtl file. The associated files should be stored in the same location as the OBJ file. The geometry data follows the header and is separated into five main categories, each with its own tag: vertices (v), texture coordinates (vt), vertex normals (vn), parameter space vertices (vp), face elements (f), and line elements (l). Vertices, texture coordinates, vertex normals, and parameter space vertices are denoted by their tag, such as "vt" for texture coordinate, and then the respective X, Y, and Z values. Face elements are denoted by their tag followed by the vertex indices that define the face. Lastly, line elements are used to represent polylines and include a list of vertices used to form the line. An example OBJ file representing a 6 in. (152.4 mm) cube can be seen in the appendix.

The AMF file

The AMF, which stands for Additive Manufacturing File Format, is an open standard for creating files to store geometry data [6]. AMF was first released in 2011 by a sub-committee of ASTM F42, which is an ASTM

(American Society for Testing and Materials) committee focused on Additive Manufacturing Technologies. The file format, which uses the file extension .amf, was created specifically for additive manufacturing and offers improvements over the STL such as support for materials, colors, lattices, and constellations. However, the underlying data is still a triangulated mesh, like that of an STL, which means that geometry limitations still exist with the AMF file. There is support for curved triangles, but this is created to minimize file size, not to increase object resolution. A curved triangle is ultimately decomposed into 1024 smaller triangles. AMF files are guaranteed to be manifold by the creation process, meaning that normal vectors are not needed in the file. All normals point outwards from the object.

ISO/ASTM 52915:2016 standard calls for the AMF to be an XML-based file. XML stands for Extensible Markup Language and defines a file format that is meant to be both human readable and machine readable. The first line of an AMF file defines the version of XML and the encoding used. Following this XML setup is the beginning AMF file tag, which surrounds the remainder of the file. Next is the header, which uses XML tags to define information about the object such as the ID, material, color, and other metadata. It is here that the units are defined. The remainder of the file is used to define the data for each of the parameters in the header. This writing will focus on the geometry data that represents the triangular mesh.

First is a vertices tag that prefaces the list of vertices, in which each vertex has its own tag. Within each vertex tag is a coordinates tag, which contains *X*, *Y*, and *Z* tags that in turn contain the actual *X*, *Y*, and *Z* values of the vertex's location in 3D space. The values of the vertex locations are defined using the units in the header. After the vertices tag is the volume tag, which contains a list of triangles, where a triangle is defined with v1, v2, and v3 tags that refer to the vertices (defined within the vertices tag) used to construct the triangle. After the list of triangles, all tags are closed and the file ends with the closing of the AMF tag, "</amf>." An example AMF file representing a 6 in. (152.4 mm) cube can be seen in the appendix.

The 3MF file

The 3MF, which stands for 3D Manufacturing Format, is an open-source file format designed and launched by the 3MF Consortium in 2015 [7]. Like the AMF, the intention of the 3MF file is to replace the outdated STL file with a new format that includes more data such as material, color, texture, and size. In addition to what is supported by the AMF, the 3MF adds support structures and multi-material capabilities. Because of all the data stored in the 3MF file, the file size can get quite large. Like the AMF, the underlying geometry data in a 3MF file is still based upon a triangulated mesh similar to the STL file. This means there are still geometry limitations for representing curved surfaces. Also like the AMF, the object is guaranteed to be manifold, so surface normals aren't needed.

The 3MF uses the extension .3mf and is an XML-based file, like the AMF. The first line of a 3MF file defines the version of XML and the encoding used. Following this XML setup is the beginning model tag which surrounds the remainder of the file. Next is the setup information, which uses XML tags to define information about the object such as the ID, base material, color, unit, and other metadata. The remainder of the file is used to define the geometry data. First is an object tag to introduce the object followed by a tag to introduce the mesh. Next is a vertices tag to begin the list of vertices, in which each vertex is its own tag that includes the *X*, *Y*, and *Z* position values of the vertex, stored using the units defined in the header. After the vertices tag is the triangles tag, which contains a list of triangles where each triangle is defined with v1, v2, and v3 values that refer to the vertices used to construct the triangle, defined in the list of vertices. After the list of triangles, all tags are closed and the file ends with the closing of the model tag, "</model>." An example 3MF file can be seen in the appendix.

Comparing the other file types to the STL

The STL file has been around, relatively unchanged, since the 1980s. Newer file options, namely the OBJ, 3MF, and AMF, have since been created and offer different functionalities from the STL, such as storage of material, texture, color, and unit information. Table 4.1 shows a comparison of various features across the file types. However, when it comes to the fundamental storage of geometry data to be used in the slicing process, all four files operate in the same way.

3D scan data from point clouds

In recent years, 3D scanning technologies have improved and their use has become more prevalent. 3D scanners work by measuring points on the surface of an object and storing those locations in 3D space. Some sensors do this with a touch probe while others will use light or laser scanning. Common practice today involves 3D scanning an object to be modified and recreated with CAD, or scanning an object to directly print a copy of it.

The scanning of an object creates a list of vertices that represent the surface of an object, much like how an STL represents the surface of an object with a list of triangles and their vertices. However, unlike the STL, the scanner data is an unorganized list of points that is stored as a point cloud

TABLE 4.1 A comparison of different geometry storage file types.

	STL (Binary)	STL (ASCII)	OBJ	AMF	3MF
Defined units	No	No	No	Yes	Yes
Color data	No	No	Yes	Yes	Yes
Material type	No	No	Yes	Yes	Yes
Multimaterial	No	No	Yes	No	Yes
Texture data	No	No	Yes	Yes	Yes
Triangulated data	Yes	Yes	Yes	Yes	Yes
Nontriangulated data	No	No	Yes	No	No
File size, 6 in cube (bytes)	684	3282	2282	1606	16,414

file [8]. Point clouds are often not watertight, as the scanning process could have missed a point or face of the object. To use a point cloud for 3D-printing, it needs to be converted into a solid mesh that can be triangulated, i.e. an STL file. Some CAD programs can import point clouds as a surface and allow the user to modify the geometry until a watertight, manifold STL can be created. For more complete point cloud files, automated programs can create a triangulated mesh and heal any small holes or imperfections in the surface so that the result is a usable STL file. Regardless of the route – manual editing or automated healing – the end result is still the same: an STL file to be used for geometry data storage and in the 3D-printing process.

Conclusion

3D-printing begins with the design process using a CAD package to create the object to be printed. The resultant object must then be exported for the slicing and printing process. The export step involves triangulating the CAD model into a mesh, typically represented with triangular faces, that stores the geometry data. The mesh, typically an STL file, is lower resolution because of the triangulation process that simplifies arcs and splines as designed within the CAD software. There are variants of the STL, such as 3MF and AMF, but the lack of rich geometry data still persists. The ensuing chapters will discuss how to make use of the STL data to slice and create toolpaths for the object. Note that, while the remainder of this book will only discuss STL data, the same principles also apply to any file type that stores geometry data as vertices and triangular faces.

References

[1] "3D CAD Design Software." Solidworks, https://www.solidworks.com/.

[2] "Your Own 3D Parametric Modeler." FreeCAD, https://www.freecadweb.org/.

[3] "Alias Software." Autodesk, https://www.autodesk.com/products/alias-products/.

[4] STL File Extension. STL File Extension - What Is an .Stl File and How Do I Open It?, 2022 https://fileinfo.com/extension/stl.

[5] OBJ File Extension. OBJ File Extension - What Is an .Obj File and How Do I Open It?, 2022, https://fileinfo.com/extension/obj#:~:text=Opening%20an%20OBJ%20file%20with%20a%20text%20editor&text=Since%20OBJ%20files%20are%20saved,text%20editor%20to%20recognize%20it.

[6] "Specification for Additive Manufacturing File Format (AMF) Version 1.2." ASTM International - Standards Worldwide, https://www.astm.org/f2915-20.html.

[7] "Specification." 3MF Consortium, 17 Nov. 2021, https://3mf.io/specification/.

[8] "The PCD (Point Cloud Data) File Format." The PCD (Point Cloud Data) File Format-Point Cloud Library 1.12.1-Dev Documentation, https://pointclouds.org/documentation/tutorials/pcd_file_format.html.

CHAPTER

5

Cross-sectioning

Michael Borish[a], *Alex Roschli*[a], *Abby Barnes*[a], *Eric MacDonald*[a,b], *Peter Wang*[a], and *Thomas Feldhausen*[a]

[a]Manufacturing Demonstration Facility, Oak Ridge National Laboratory, Knoxville, TN, United States, [b]University of Texas at El Paso, El Paso, TX, United States

Abstract

Now that an object has been transformed to an STL representation, the first major step of the slicing process can begin: cross-sectioning. This step converts a 3D object into a stack of polygons. Polygons are closed, 2D shapes composed of straight edges. These polygons are typically the result of intersecting a plane with an object at various heights along the z-axis. These stacked polygons create the "layers" that are characteristic of the 3D-printing process. In addition to the cross-section operation itself, several details need to be addressed. These details include necessary preprocessing and potential stitching, smoothing, and simplification of the cross-sections.

Introduction

With an STL representation of an object in place, the slicing process can begin. In principle, the cross-sectioning of a faceted representation of an object with a plane is very straightforward. The devil, as the saying goes, is in the details. However, before discussing the details of cross-sectioning, it is necessary to describe the typical preprocessing that takes place. This preprocessing can take place at various points prior to beginning the slicing process. In general, most slicers perform necessary preprocessing when the STL is loaded into the software package. While preprocessing computations could be included at the start of the cross-sectioning process, the result of preprocessing is static and geometry-dependent. It therefore only needs to be computed once, so it makes more sense to

https://doi.org/10.1016/B978-0-443-15286-3.00011-4

complete pre-processing separately from and prior to cross-sectioning, which the user may decide to do multiple times.

Pre-processing

Recall from *Chapter 4: Geometry data storage* that an STL file provides the minimal necessary information to construct a solid body. An STL file is a faceted representation of a solid body composed of a sequence of vertices. Each set of three vertices describes a face and each face may be associated with a normal that helps determine the "outside" of the object. No other information is provided, and this is why STLs are considered to be "polygon soup" when attempting to render or manipulate them. For many geometric processing tasks, additional information about the connectivity and ordering of vertices is required. This information is generated by preprocessing. There are many different data structures that can represent this information and the amount of information gathered from preprocessing can vary. At a minimum, for traditional slicing, it is critical to have the following pieces of information: the face that every vertex belongs to, all the vertices that construct a face, and the neighboring face that a vertex touches.

To those readers who have a background in graphics or computer science, this additional information should sound like a very old and well-known data structure. That structure is the half-edge [1]. A full description of the data structure is outside the scope of this book, and readers who are interested can easily find information with a cursory internet search. In principle, the half-edge seeks to allow easy navigation of unordered polygons by connecting vertex and face information with that of neighboring faces. The half-edge is generally regarded as striking a balance between complexity and the underlying represented data. A simple example is shown in Fig. 5.1.

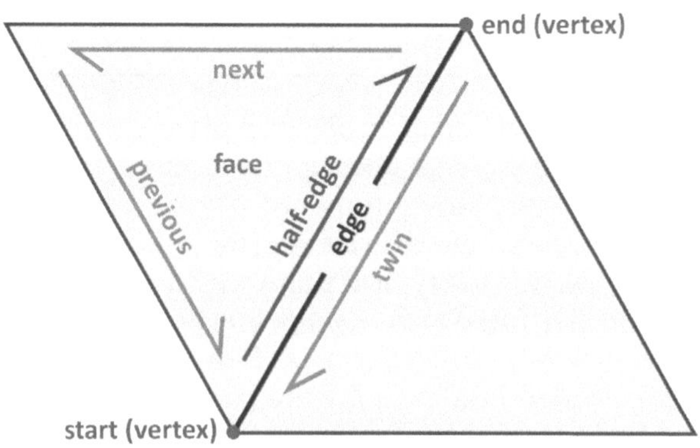

FIG. 5.1 Half edge data structure. Each vertex has connectivity information to walk its own face as well as the neighboring face of a triangulated object.

From this data structure, or something similar, navigation becomes more straightforward. For example, an iteration can begin at any vertex. From that vertex, the associated edge is known, and from that edge, the neighboring edge can also be determined. This allows an algorithm to then "walk" through all of the vertices by continuously moving to neighboring vertices. With this connectivity information in place, the cross-sectioning step can begin.

Cross-sectioning

The most typical cross-section of an object in slicing software is by a horizontal plane. The plane is incrementally moved up the z axis, according to user settings for the layer height. If the layer height is h and the number of layers is n, an object is $n \times h$ tall. How tall is $n \times h$ in terms of real-world distance? That depends on two factors: the units for the STL and the units for the slicing plane. Remember, an STL is unitless. Typically, slicing packages assume a specific unit for all imported objects. For example, all STLs are assumed to be in millimeters. Similarly, in most slicing software packages, when a user defines their layer height, it can be in any unit, such as inches or millimeters. One of the challenges of all steps of the slicing process is to interpret these preferred and human-readable units in a way that makes computation of the final pathing achievable. One way of accomplishing this is the use of high-accuracy representations to combat numerical inaccuracy. For example, if all the computation inside the slicing package is accomplished in microns, then having users describe heights and distances in terms of inches or millimeters means that any numerical inaccuracy will likely be truncated when the final pathing is produced.

To create a 2D cross-section, each of the faces of the object must be evaluated in 3D Cartesian space (using XYZ axes). When cross-sectioning with a simple horizontal plane, all that is necessary is to compare the Z component of each of the vertices of each face to the layer height in question. Such a comparison results in eight possible outcomes: six valid outcomes that generate segments that will be assembled into a polygon, and two degenerate outcomes that do not produce segments. The six valid outcomes arise simply because of permutations of the orientation of the face and its associated vertices, as shown in Fig. 5.2.

Fig. 5.2 also shows how each face is evaluated when using a horizontal plane of intersection, Z_c. During cross-sectioning, this plane is moved up the Z axis in intervals equal to the layer height. As the plane moves, the vertices V_0, V_1, and V_2 of each of the triangles in the STL are tested against the current position of Z_c. If the plane intersects a triangle for any of the cases shown, the intersection points I_0 and I_1 will be created just

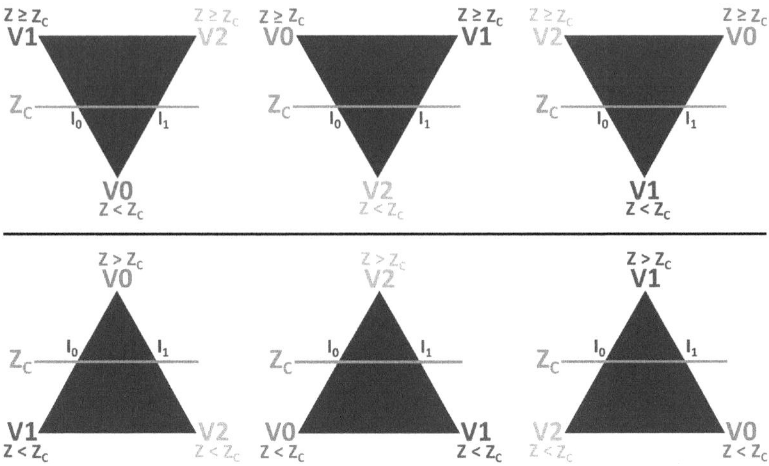

FIG. 5.2 Intersection of a horizontal plane with a triangle of the STL. Z_c represents the height of the cross-sectioning plane, I_0 and I_1 the intersection points, and V_0, V_1, and V_2 the vertices of the face.

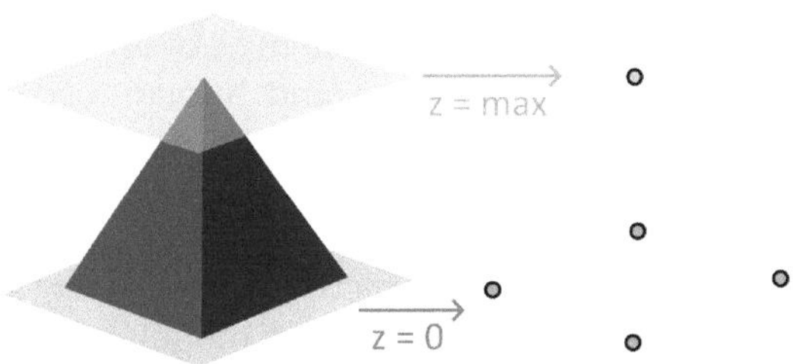

FIG. 5.3 Two degenerate cross-sectioning cases. (1) Z = max: results in a single point with no area. (2) Z = 0: results in multiple points without connectivity because the entire polygon is contained within the intersection plane. Again, this results in no area.

as shown in Fig. 5.2. Once the cross-sectioning process is complete, all of the resulting intersection points and the line segments between them form the perimeters of the resulting 2D polygonal layers.

As previously mentioned, there are some intersections that do not form valid vertices. These two degenerate cases are easily demonstrated using a pyramid like the one in Fig. 5.3. Consider a horizontal plane intersecting the pyramid at height $Z=0$ and height $Z=Z_{max}$, where Z_{max} is the vertex with the largest Z value. At a height of $Z=Z_{max}$, the intersection of this object will be a point. A point has no area that any reasonable pathing can fill and so it is discarded. At the height of $Z=0$, the situation is very similar. In this object, the polygon of the base of the pyramid is entirely parallel to, and contained within, the slicing plane. Therefore, almost all the edges are also parallel to the slicing plane and cannot form intersections. The only intersections occur at the corners, which results in four singular points. This is the same situation as Z = max; a set of points with no area or connectivity that are discarded. Put another way, all the edges and their half-edge companions both exist in parallel to the slicing plane and, as such, do not produce valid segments that can be used to construct a closed polygon.

Due to these degenerate cases, each layered slice is offset by half of its layer height. For our $n \times h$ tall object, each layer will be sliced at $n \times (0.5 \times h)$, $n \times (1.5 \times h)$, and so on. Doing so removes the possibility of encountering these degenerate cases at the bottom and top of an object. More importantly, this offset also more accurately captures the cross-sectional geometry that will ultimately be constructed. Remember, these cross-sections represent a real-world height. As a result, when a 3D printer constructs an object, its nozzle or build point will not be at Z = 0 when the build begins. This is especially true for extrusion-based systems because there must be room for the material to exit the nozzle.

From the valid cases, two points of intersection can be calculated for each face, and a new segment is generated between those two points. Using the first picture from Fig. 5.2, for example, the intersection points are midway between V_0–V_1 and V_0–V_2 and are represented as I_0 and I_1. A segment connecting I_0 and I_1 is drawn across the surface of its face and is bundled with the previously described connectivity information. Once all faces have been evaluated, the results are forwarded to the next major stage of the cross-sectioning step.

In the next stage, the newly generated segments need to be ordered and connected. Recall that an STL for 3D-printing should be manifold and watertight. Such a requirement guarantees that a boundary polygon can be generated by walking the segments of the intersected triangles along the surface. So, how is this accomplished? Remember the connectivity information of the half edge, shown in Fig. 5.1. Each of the generated segments is the result of evaluating the vertices of a face. Knowing the vertices means that the edge those vertices form is also known, and therefore the neighboring faces are known as well, because of the relationship between the half-edge and its twin. It is therefore straightforward to move from the first intersection point on a face (I_0) to the second intersection point of that face (I_1) and then to the neighboring face that shares the second point's edge (i.e., has its half-edge). These steps are repeated for all faces intersected by the plane. The result of this process is a fully connected polygon defining the solid volume for which pathing will subsequently need to be generated. This polygon is called the layer polygon.

Every layer polygon will have an imposed ordering: clockwise or counterclockwise. This order describes the direction of the evaluation of each face—how I_0 and I_1 are oriented, or in other words, whether a face's second intersection point is to the left or right of the first. The order is imposed by the way the STL's faces are defined. Typically, the vertices of each face of an STL are specified counterclockwise, with the normal facing "outward." A polygon that represents solid areas will therefore have its vertices ordered counterclockwise, while polygons that represent holes will have vertices ordered clockwise. This is known as the winding order of the polygon, which will be discussed more in the next chapter.

At a basic level, that is all that is required to generate a closed polygon. It is important to note that, although the discussion has focused on evaluation of the Z component of the triangles to generate a polygon, in many cases, the Z components of all the vertices of the resulting polygon are dropped. This is to speed computation and is acceptable because the polygon and all its pathing will have the same Z coordinate as the respective layer height. As a result, the points in the subsequent pathing do not have a Z coordinate because it is unnecessary. However, many advanced pathing approaches, such as off-axis and nonplanar solutions, require the Z component because the process is no longer limited to a horizontal plane. Therefore, in general, it's a good idea to maintain the Z component information, but not use it during computations.

Once a closed polygon is generated, the slicing process can move to the next step of the process: closed-contour path generation. However, not all STLs are appropriately modeled and not all mathematical computation will proceed exactly as planned. Further, while the polygon generated may be mathematically correct, it is important to remember that these algorithms are an approximation for a real-world construction process. As such, considerations for the construction process must be superposed onto the polygon that the cross-sectioning creates, to help ensure that the polygon is suitable for 3D-printing. At this point, the details that were previously mentioned come into play.

Optimizing polygons for 3D-printing

Recall from *Chapter 4: Geometry data storage* that non-watertight STL's need to be "stitched." Stitching is one polygon optimization that is often done as part of the polygon generation process, while the newly identified segments of-the faces are being "walked." Remember, these segments are the result of the intersection of the plane with a triangular face. The start and end points exist somewhere along the edges between the vertices of a face. The most common issue occurs when neighboring faces are sliced into two segments that should share a common point—one at the end, the other at the start—but do not. Instead, the end of the first segment does not perfectly match the location of the next segment's start point. This is most commonly the result of numerical inaccuracy inherent in all mathematical computation. The typical solution is to allow a fuzzy comparison. Two points are considered identical if they are within a specified distance from one another. For example, if two points are within a few microns of one another, they are considered identical, and the polygon construction process continues.

A slightly more significant issue arises when the end and start points are a significant distance from one another, such that this small tolerance is insufficient for connection. At this point, the remedy is less clear. The exact issues and resolutions form the foundation for the mesh healing techniques mentioned in the previous chapter. These techniques can be applied as part of preprocessing or cross-sectioning. In general, they attempt to fix issues with the 3D object that would otherwise result in mismatched points. This subject area is expansive and outside the scope of this book, but the following subsections will discuss some examples and high-level remedies.

Unclosed polygons

Suppose a face or small handful of faces are disconnected or inverted. This could be the result of an inverted normal vector for a face, or a seam that wasn't properly closed when the object was designed in CAD. Regardless, when cross-sectioning this object, segments of the contour of the resulting polygon are disconnected; they do not form a closed polygon. There will be a hole or small number of holes. The problem is a classic one: how to join a set of sequences of connected points, called polylines. If only one segment is missing or disjointed, then the set only contains one polyline, and the solution is straightforward: simply close the start point and end point. The solution is less obvious for a set containing multiple polylines, as is the case if multiple segments are missing or disjointed.

One solution is to use a sparse grid representation. A sparse grid allows the efficient representation of high dimensional data. In this case, the grid represents the start and end points of the polylines in question. By tracking these locations in a sparse grid, possible stitch locations can be identified between the start and end points of the set of polylines. The possible combinations can then be tested to identify the stitches that produce the least amount of error, where error is usually defined as the total distance to be stitched. This ensures that stitches are formed between the start and end points that are closest to each other. Then, those start and end points are joined with individual polylines, which are reversed as necessary to ultimately form a closed polygon.

Overlapping segments and collinear points

Another modification to the boundary polygon that may need to be applied is the removal of vertices from overlapping segments or nearly collinear points. These are easy simplifications that improve the overall quality of the printing process, especially for large format 3D printers. Consider the case of the simple cube shown in Fig. 5.4. The cross section of this cube is a square, but the square will be composed of eight points because each surface of the cube is composed of two triangular faces. If these points remain in the cross-section, the pathing that is ultimately produced will contain eight unique paths, even though four would be sufficient. While there is nothing inherently wrong with this representation, issues may arise when considering the hardware of the system in question. Most motion controllers will command the machine to accelerate and decelerate when approaching a start or end point. If the square has eight paths, then the beads that are ultimately constructed will be non-uniform at the midpoint due to this additional acceleration and deceleration. The preferred alternative would be to remove the superfluous vertices and have four uniform beads. This is an issue not just for truly collinear points, but points that are nearly collinear as well. The ultimate tolerance for collinearity will be a combination of factors including the material, the size and type of build process, and requirements of object quality.

Smoothing

The final issue to discuss is smoothing. Recall from the previous chapter that STLs are faceted representations. Because of this, objects with curvature are approximated by some number of straight-line segments. An example was shown in Fig. 4.2 of *Chapter 4: Geometry data storage*. High resolution STLs will have many segments to closely approximate the curve. Just as with collinear points, there is nothing inherently wrong with this representation, but similar issues can arise. Many small segments will

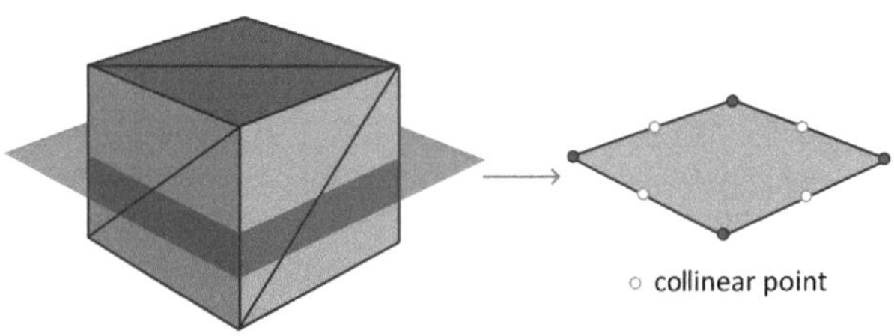

FIG. 5.4 Example of collinear points in an STL. The removal of the collinear points results in a simplified and improved representation of the cross-section.

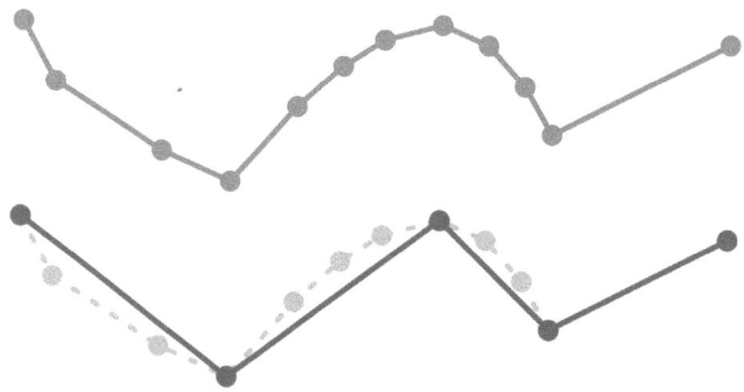

FIG. 5.5 Example of Douglas-Peucker smoothing algorithm. The top polyline represents the initial input while the bottom polyline shows the simplified output. Just like smoothing, this process can result in better-quality 3D prints.

produce many small, jerky motions as the 3D printer attempts to follow all the pathing. This is an issue, particularly for large format systems because the system has to move a significant mass. In this case, once the pathing process has been complete, it is generally a good idea to sacrifice a small amount of accuracy in favor of smoother motion. This is accomplished using smoothing algorithms that remove some of the vertices. There are multiple algorithms that address this issue; one of note is Ramer–Douglas–Peucker or simply Douglas-Peucker [2,3].

The Douglas–Peucker algorithm is an iterative end-point fit algorithm that simplifies a curve composed of line segments into a similar curve of fewer points. Though it was originally intended for cartographic simplification, it works well here. The algorithm iteratively evaluates the points on a polyline. If a point is within the error tolerance, it is removed from the polyline, otherwise it is kept. Error is typically represented by some type of distance calculation such as the Hausdorff distance between curves. When all points have been evaluated, a simplified curve is produced. An example is shown in Fig. 5.5.

The tolerance of this algorithm must be balanced with the information lost. Too many segments in the cross-section will cause jerky motions during construction, but too much smoothing will decrease the accuracy of the build. Just like the situation with nearly collinear points, the tolerance will be dependent on a number of factors that vary by situation.

Conclusion

Cross-sectioning is one of the first, fundamental steps of the slicing process. This chapter has introduced the basics of cross-sectioning of objects for 3D-printing. Even though the process is straightforward, there are several common issues that can arise and need to be addressed by stitching, smoothing, and simplification. In addition to highlighting these issues, this chapter has also discussed their resolution through some of the basic solutions that are usually employed. The processes described in this chapter form the basis for more complex approaches that are being used for next-generation pathing solutions, which will be touched on in later chapters.

References

[1] B.G. Baumgart, A polyhedron representation for computer vision, in: Proc. AFIPS Natl. Comput. Conf, vol. 44, AFIPS Press, Alrington, VA, 1975, pp. 589–596.

[2] U. Ramer, An iterative procedure for the polygonal approximation of plane curves, Comput. Graph. Image Process. 1 (3) (1972) 244–256, https://doi.org/10.1016/S0146-664X(72)80017-0.

[3] D. Douglas, T. Peucker, Algorithms for the reduction of the number of points required to represent a digitized line or its caricature, Can. Cartogr. 10 (2) (1973) 112–122, https://doi.org/10.3138/FM57-6770-U75U-7727.

CHAPTER

6

Closed-loop toolpath generation

Alex Roschli[a], *Michael Borish*[a], *Abby Barnes*[a], *Thomas Feldhausen*[a], *Eric MacDonald*[a,b], *and Peter Wang*[a]

[a]Manufacturing Demonstration Facility, Oak Ridge National Laboratory, Knoxville, TN, United States, [b]University of Texas at El Paso, El Paso, TX, United States

Abstract

Closed-loop paths, also known as closed-loop contours, are paths that start and stop at the same point and are typically used to define the perimeter or outermost boundary of the polygon that represents a given layer of an object, referred to as the "layer polygon." Finding the location of a closed-loop path involves applying two offsetting steps to the edge of the layer polygon. The offsetting steps respectively serve to (1) find the area that contains the toolpath and (2) find the centerline of that area, to define the path itself. After finding the perimeter path, the remaining geometry from the layer polygon can be used for additional path-planning steps. This chapter will discuss how the location for a closed-loop path is found, including various problem scenarios, and the different types of closed-loop paths used in additive manufacturing.

Introduction

Once an object has been loaded in a slicing program, cross-sectioned, and all layer polygons have been generated, the process of fitting toolpaths can begin. There are three main types of toolpaths: closed-loop, open-loop, and space-filling [1,2]. A closed-loop toolpath is referred to as such because its start and end points share the same coordinates, i.e., it makes a complete loop [3]. This is in contrast to the start and end points of an open-loop toolpath, which have distinct coordinates [4]. Space-filling toolpaths can be open- or closed-loop but are distinguished by the fact that they are typically used to fill areas, instead of contour them [5].

https://doi.org/10.1016/B978-0-443-15286-3.00012-6

This chapter focuses on closed-loop paths, which are typically the first type of toolpath to be generated and fit to a layer because they often represent the boundary or contour of the layer polygon. To ensure no other paths disrupt the boundary of the layer polygon, the boundary must be pathed first. The following sections will cover the fundamental concepts of closed-loop path generation, including all the steps needed to find closed-loop paths and the steps needed to fit them to a layer. Later chapters may discuss different considerations, modifiers and optimizations that can affect the closed-loop contours presented here, but the fundamental concepts will remain the same.

Finding the path location

The most important aspect of planning any toolpath is defining its location. This includes finding the area to be filled by the path as well as defining the centerline that the path will travel to fill that area. To find the area for any toolpath, the user must first define the desired bead width (also known as the extrusion width, track width, or road width) of the toolpath. This bead width determines the overall width of the printed bead but does consider any overlap with additional beads. While the 3D-printing process does not inherently prevent the bead width from changing during a toolpath, the computations required to achieve this with any degree of accuracy are very complex and are absent from most, if not all, slicing programs. This chapter will therefore focus on planning a closed-loop path that has a consistent bead width for its entire length.

The procedure of finding a closed-loop toolpath's location involves two offsetting steps, one inward and one outward. The inward offsetting step is done first in a process called erosion. During erosion, the boundary of the layer polygon generated during cross-sectioning is offset inward by the bead width. The new polygon, the area that remains after offsetting by the bead width, is the remaining layer polygon, which will be saved for additional path planning. Fig. 6.1 shows a rectangular object being offset inward (eroded) by one bead width.

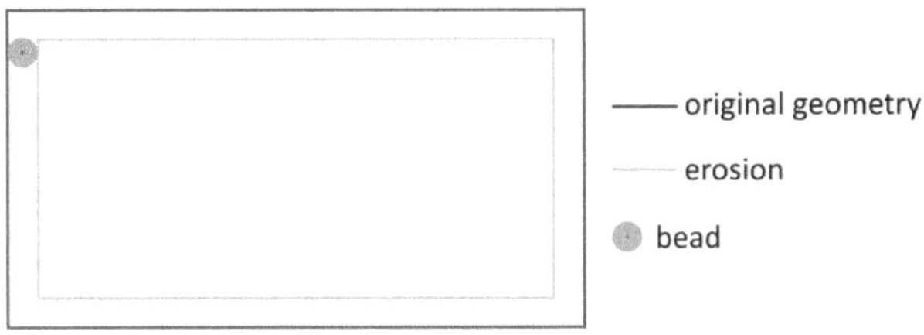

FIG. 6.1 Layer boundary and the result of the erosion offset by a full bead width.

To find the centerline—the location where the toolpath will be planned—the remaining layer polygon is offset outwards by half of the bead width in a process called dilation. This offset outward by one half width creates a new polygon whose boundary represents the centerline through the area between the original layer polygon and the remaining layer polygon. The points that make up the polygon will be the points used in future steps for adding path modifiers and writing g-code. Fig. 6.2 shows the centerline location found after the outward offset (dilation) by one half bead width.

An obvious question arises when looking at the resulting centerline: why two steps are needed? Why not just start from the original layer polygon and offset inwards by one half bead width to define the centerline of the toolpath? The reason this approach doesn't work is because a single offset operation, using only half of the bead width, doesn't account for the area covered by the other half of the bead. Take for example a compound shape made up of a square with a small, 1.5-bead-wide rectangle attached to one corner (Fig. 6.3). Offsetting by one half bead width creates a centerline all the way around the object (Fig. 6.3A, green line) that does not overlap or intersect itself at any point. However, the centerline of each long edge of the rectangle gets within half of a bead width of the centerline of the other long edge. Printing one of the long edges will

FIG. 6.2 After the erosion step in Fig. 6.1, the dilation step is used to find the toolpath centerline (dashed line).

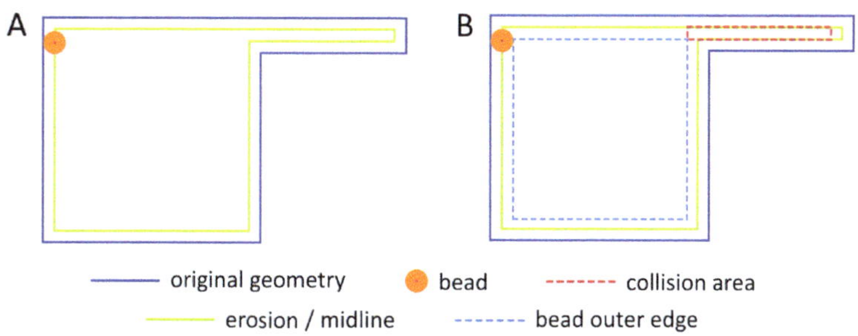

FIG. 6.3 An object where a half bead width erosion (A; green line) doesn't cause overlap, but a full bead width erosion (B; dashed line) does.

not create an issue, but when printing the second long edge, whose pathing is within half a bead width of the first bead's centerline, a collision will occur (Fig. 6.3B). This is because the edge of the second bead meets the centerline of the first bead; there is only enough space for half of the second bead.

If this path is planned by offsetting in two steps, inward by one full bead width and then outward by one half width, the outcome is different. The inward offset by one width causes the centerline to cross over itself within the area of the rectangular portion of the shape (Fig. 6.4A). Any areas of overlap become negative area and are removed from the polygon. The negative area shown in Fig. 6.4 gets deleted, leaving just a square-shaped polygon for the remaining layer polygon. Then, when this polygon is offset outward, in the dilation step, a larger square shape is created, and the resultant pathing is just for the square (Fig. 6.4B). The area of the rectangle portion does not get printed. The inability to use a closed-loop toolpath to successfully print a geometry with any dimension under two bead widths will be discussed further in the next section.

After a closed-loop path has been found and the path location has been planned, the remaining area of the layer that is unplanned is saved and passed to the next step to plan additional paths. To find all of this unplanned area, a polygon comparison is used to subtract the area of the planned paths from the original layer polygon. The next step, planning additional paths, can include closed-loop paths, in which case this process of erosion and dilation begins again, starting with the remaining layer polygon. The next step could alternatively be open-loop pathing or space-filling toolpathing, or the object could be left hollow and have no additional paths planned for the layer.

It is important to note that central to the process of erosion and dilation is the winding order of the polygon that is being evaluated. Recall from *Chapter 5: Cross-sectioning* that cross-sectioning occurs on the "triangle

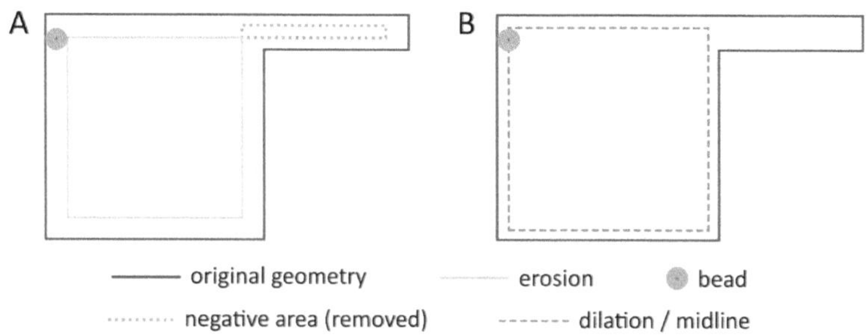

FIG. 6.4 The pathing when a full bead width erosion (A) is applied results in just a square shaped toolpath (B; dashed line).

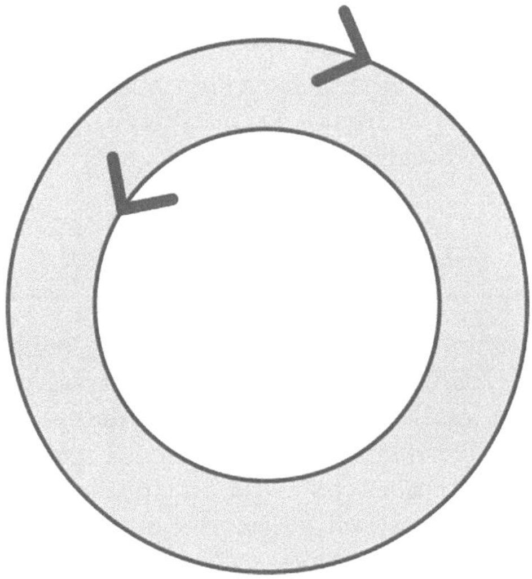

FIG. 6.5 A doughnut shape where the outer polygon is wound counterclockwise, and the inner polygon is wound clockwise.

soup" of an STL. The resulting polygons, though, have a specific order of vertices, termed their "winding order," that is imposed as part of the polygon creation process. The order of a polygon's vertices dictates whether the polygon erodes inwards, towards the centroid of the polygon, or outwards away from the polygon's centroid. Dilations then always occur in the opposite direction (away from or towards the centroid) of the erosion step. Typically, erosions move inward for polylines that represent a solid, closed geometry, which have a counterclockwise winding order. Erosions move outwards for polylines that represent holes, which have a clockwise winding order. These orderings are not set in stone though. The important aspect about winding order is that polygons that represent solid areas vs holes have different winding orders, to allow differentiation of any number of nested polygons. Fig. 6.5 shows the difference in these winding orders.

As this suggests, at a high level, the process of erosion and dilation creates closed-loop contours by moving all line segments toward or away from the centroid of the polygon. At each step of this process, line segments that overlap or have gaps between them are either clipped or stitched together, respectively. For this reason, offsetting cannot be applied to polylines in the same way, because polylines may be open and have no centroid. Additionally, the erosion-dilation process may be applied to geometries with arcs, but higher order polynomials require a different approach that is briefly mentioned in *Chapter 18: The G-code file.* Like line segments, the erosion and dilation of an arc both preserve the arc's shape. The erosion of a spline, however, can alter the shape of the spline if the offset causes points of the spline to cross over each other. The region after the intersection would become negative area and be removed, and this change to the shape would be carried over to the dilation process.

Two-bead-width constraint

As previously mentioned, the nature of the closed-loop offsetting process prevents closed-loop paths from being used to print objects that have at least one dimension that is less than two bead widths, if the bead width is kept constant. When an object like this is sliced, the closed-loop pathing process does the initial offset inward by one bead width, causing the boundary of the resultant polygon to overlap with itself and have negative area (Fig. 6.6). The negative area is removed, which eliminates all of the polygon. The result has no polygon to dilate, or offset outwards, so no path can be planned. This is known as the "two-bead-width constraint," wherein objects to be printed are preferably designed so that both their width and length (and those of any protrusions) are greater than or equal to two bead widths.

Using a consistent bead width, any wall that is less than two-bead-widths wide will not have pathing. Similarly, any wall that is at least two-bead-widths wide but less than 3.99-bead-widths wide can only have a single closed-loop path. Fig. 6.7 shows three different wall widths and the resultant pathing.

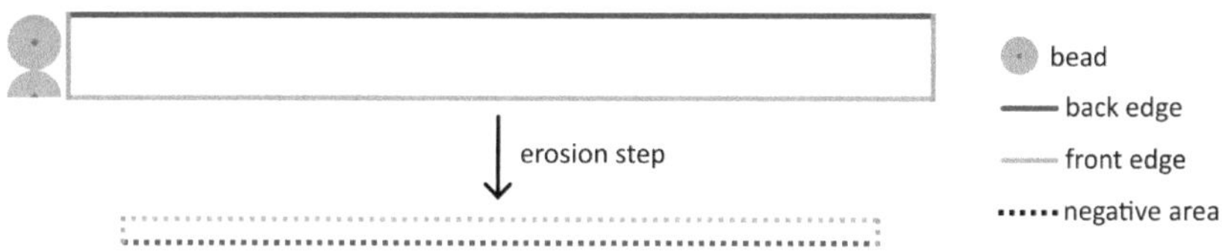

FIG. 6.6 A 1.5 bead width wall causes overlap and negative area during the erosion offsetting step.

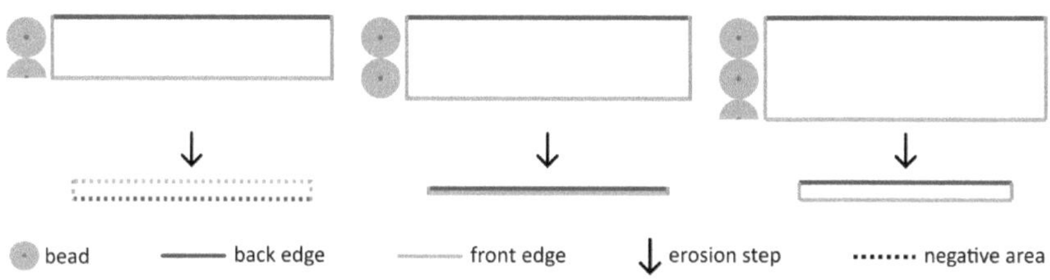

FIG. 6.7 The resultant pathing in three scenarios: 1.5 bead width wall, 2 bead width wall, 2.5 bead width wall (from left to right). No pathing is created for the 1.5 bead width wall.

Multiples of the bead width

The two-bead-width constraint requires a polygon to be at least two-bead-widths wide in order for a single closed-loop path to be fit to the polygon. At the beginning of path planning for a layer, a closed-loop path can be planned as long as the two-bead-width constraint has been followed. After that closed-loop path is planned, the remaining layer polygon can be fit with additional closed-loop paths, but the two-bead-width constraint must be applied to every new remaining layer polygon. For example, after a second closed-loop path is generated, a new remaining layer polygon is found that excludes the area covered by the first and second closed-loops. As long as the new polygon is at least two bead widths wide, a third closed-loop path can be planned. This process can be used to continually generate closed-loop paths until the remaining polygon does not meet the requirements for the two-bead-width constraint, or until there is no polygon remaining.

The implication of this is that, in order to be planned entirely with closed-loop paths, a layer polygon must be an even-integer multiple of the bead width. An object that is five-bead-widths wide (Fig. 6.8) can be fitted with two closed-loop paths, taking up a total of four bead widths, but a third will fail the two-bead-width constraint because there isn't space for six bead widths. The space remaining after the second closed-loop path is planned can be filled with additional pathing types, which will be discussed in future chapters.

The closed-loop path generation process can be modified to allow for a small amount of error, or a small amount of overlap, to occur, such that an object that is 1.9999 bead widths wide would still be fit with one single closed-loop path. The amount of error, or overlap, can usually be configured as a user setting within the slicing program. Alternatively, the slicing parameters can be adjusted such that the width of a bead is smaller than one. For example, a 1.8-bead-widths wide object typically wouldn't generate a single closed-loop path, but if the bead width is set to 0.9, a closed-loop path will be generated.

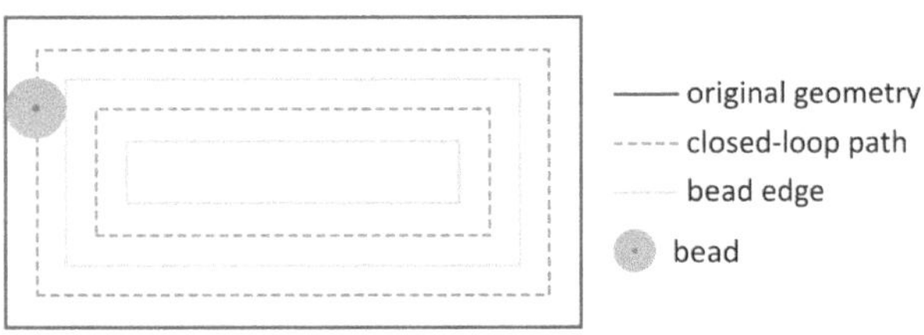

FIG. 6.8 A box that is five bead widths wide will result in two closed-loop paths and a void space in the middle.

Hollow shapes

For hollow objects such as a doughnut shape—a large circle containing a smaller concentric circle—the erosion step creates offsets from the outer and inner boundary simultaneously. This causes the first step of creating a closed-loop toolpath to result in two separate closed loops, one around each boundary of the object. This means that the two-bead-width constraint applies such that the width of the doughnut, the distance between the outer and inner circles, must be at least two bead widths to produce the geometry. Fig. 6.9 shows the geometry and toolpaths of a 2-bead-width doughnut. If the doughnut is designed as just one single bead width, the geometry isn't printed because no toolpath is formed, as shown in Fig. 6.10.

In this case, the two-bead-width constraint can be circumvented to print hollow, single bead width objects with just a single closed-loop path. To do this, the object must be designed as a solid—in this case, a single circle rather than a doughnut. The user can designate just one closed-loop path to be generated, meaning the erosion and dilation processes happen

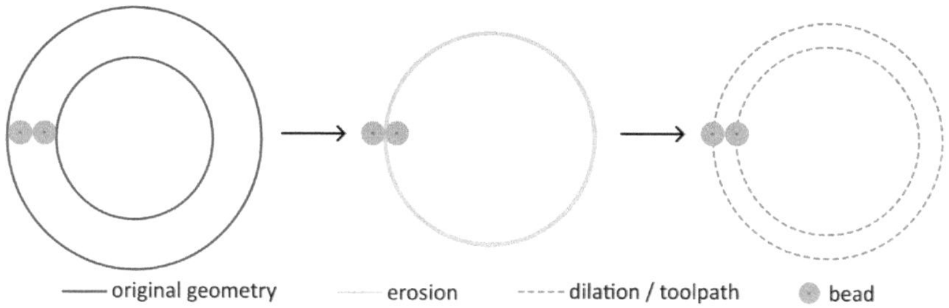

FIG. 6.9 A two bead width doughnut can be fit with two closed-loop toolpaths.

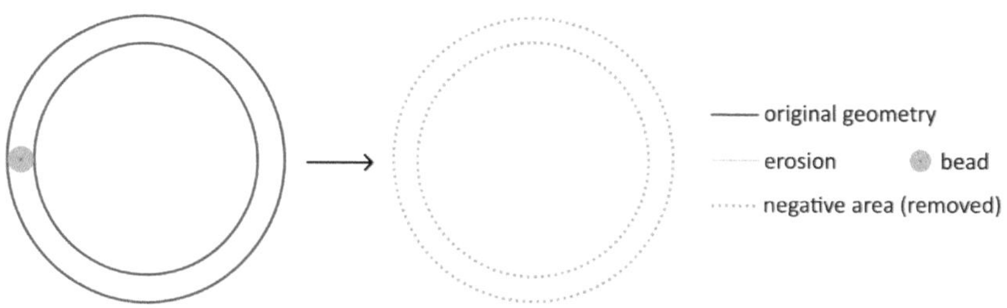

FIG. 6.10 A one bead width doughnut cannot be fit with closed-loop toolpaths because the erosion step causes negative area.

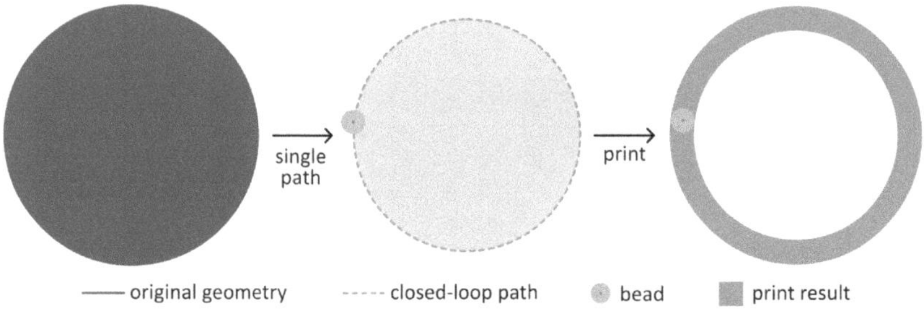

FIG. 6.11 A solid circle and the resultant pathing after one closed-loop path is generated.

just once, for the outer boundary created by the circle. After this closed-loop path is found, the remaining geometry is passed to the additional pathing steps, but if those path types are disabled, the resultant layer tool-path will be just one closed loop. Thus, a hollow object will be constructed. This can be seen in Fig. 6.11.

Edge cases

When it comes to layer polygons and offsetting, it's natural for edge cases to occur that the slicer must handle. For example, take a solid compound shape made up of two overlapping circles, like the one shown in Fig. 6.12A. The result of generating one closed-loop path is a path that follows the entire contour of the object. However, after the first closed loop is generated, the remaining layer polygon contains two separate polygons. These are called islands. If the slicing has been configured to generate two closed-loop paths, a second closed loop will be generated for each of the islands in the remaining layer polygon (Fig. 6.12B).

If the dimensions of the shape from Fig. 6.12 are modified slightly such that the connecting area has a width less than two bead widths, the edges of the connecting area will overlap when the shape is eroded. They will then be removed, and the connection will be broken. This ultimately creates two separate lobe-shaped islands for the layer. During the dilation step, each polygon will be offset outwards, creating paths around each lobe. This type of shape and the resultant pathing is shown in Fig. 6.13.

Importantly, the pathing for each island is generated independently. That way, if one island can support a second closed-loop path, but the other cannot because its remaining layer polygon violates the two-bead-width

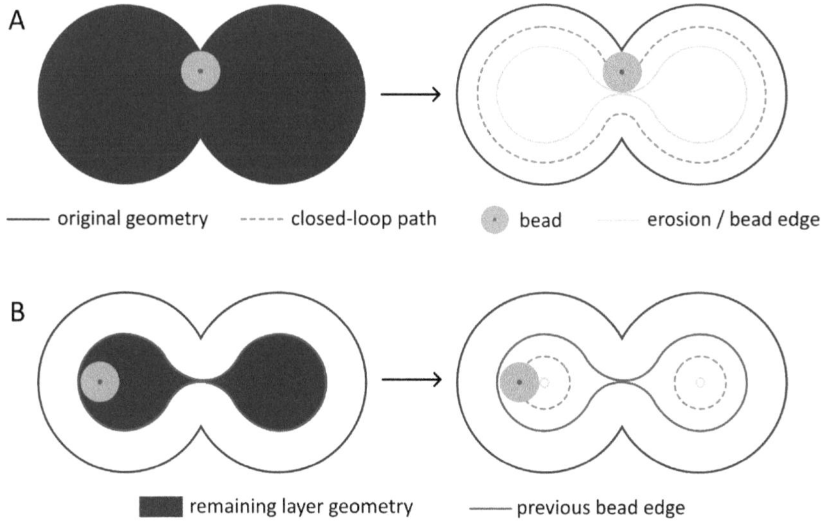

FIG. 6.12 (A) Generating one closed-loop path around a figure eight shape; (B) when adding the second closed-loop path, there is not space for a single path around the whole boundary, and thus two smaller paths are generated around each polygon's boundary.

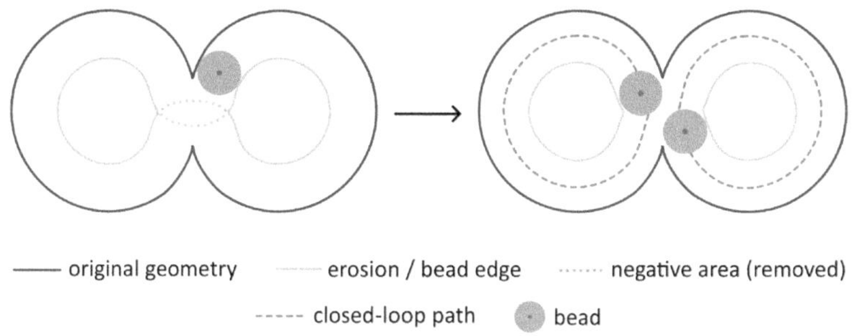

FIG. 6.13 The center of this figure eight is not two bead widths wide and thus cannot support one single, closed loop around the entire boundary of the layer.

constraint, one island will have two closed-loop paths while the other will have just one. For example, the compound shape shown in Fig. 6.14 is made up of two overlapping circles like the one in Fig. 6.12, but the circles are unevenly sized. Like the shape in Fig. 6.13, the edges of the connecting area will overlap and be removed during the erosion step, and the two circles will become separate lobes. In this example, however, the lobe on the right is fitted with a second closed-loop path, but the lobe on the left is not.

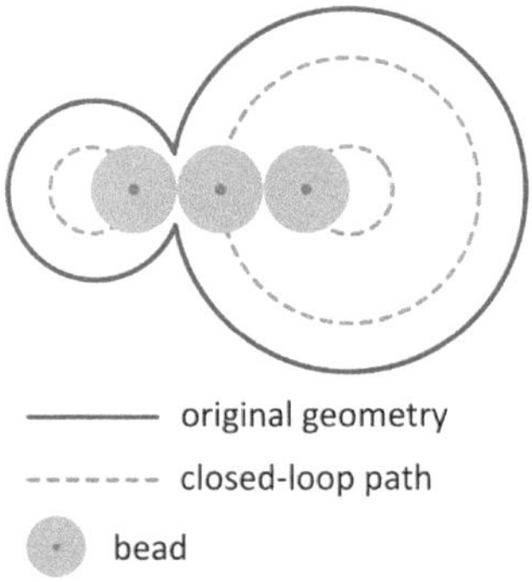

FIG. 6.14 After the first closed loop path is generated, creating two independent closed-loop paths (one around each lobe), a second path won't fit around the left lobe but will fit around the right lobe.

Complex path geometry

For many geometries, the default shape and layout of pathing is counter-intuitive and often results in excessive pathing, which is not ideal when trying to optimize print time and material usage. Take for example a box with an interior truss structure, where all segments are two bead widths wide (Fig. 6.15). Most people would draw the rectangular perimeter first and then add the truss in the center, so that they only have to pick up their pen or pencil one time. However, that is not how a computer calculates the pathing using the erosion and dilation process previously discussed. The algorithm works by offsetting the entire polygon by one bead width. After dilating by half a bead width, the result is many separate paths.

To generate pathing that is just two outer contours and the interior truss, the part geometry can be slightly modified to force the computer to "think" like a person. The only modification needed is a separation of the truss from the outer rectangle, which can be achieved by placing small cuts along the truss where it would normally join to the outer structure (Fig. 6.16).

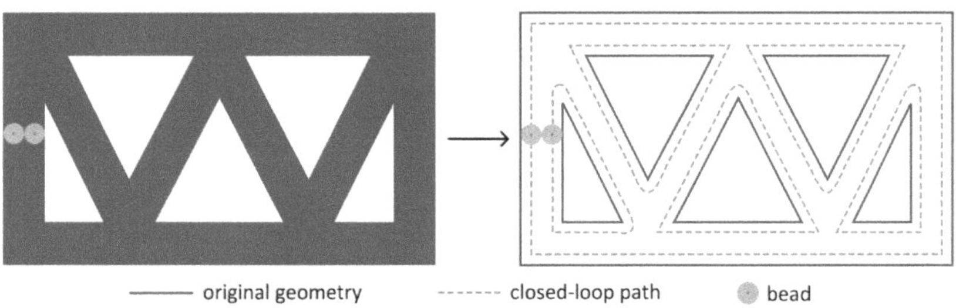

FIG. 6.15 A two bead rectangle with interior truss structure (left), and the resultant pathing for the structure (right).

FIG. 6.16 The same truss from Fig. 6.15, but with cuts added to separate the truss from the outer rectangle.

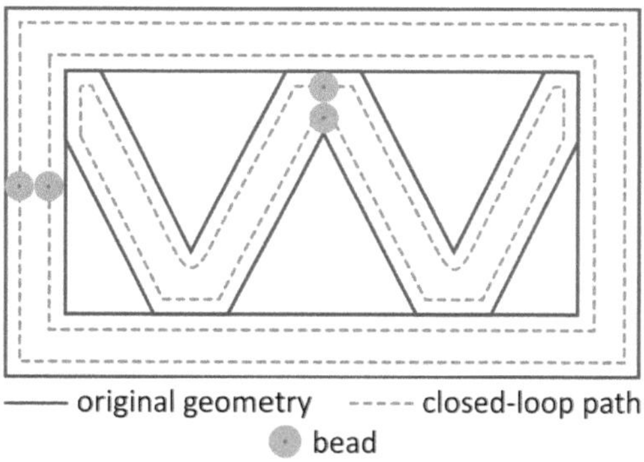

FIG. 6.17 The pathing for the object in Fig. 6.16.

The original shape from Fig. 6.15 is represented using a single polygon for the layer. The modified shape, from Fig. 6.16, is represented with two polygons: one for the outer rectangle and one for the truss. These function like the islands discussed previously; during closed-loop path generation, each polygon has pathing generated independently. This results in two closed-loop outer contours for the outer, rectangular structure and one closed-loop path for the interior truss (Fig. 6.17). If the cuts from Fig. 6.16 are made very small (a width of less than 1% of the bead width), the material flow of the 3D printer should allow the inner truss to still bond to the outer rectangle and not leave a gap in the part geometry.

Types of closed-loop contours

There are two main types of closed-loop contours used for toolpathing: perimeters and insets. "Perimeter" is typically the term given to the outermost closed-loop path on a layer, as this is where the perimeter of the layer is. This can be an outer perimeter, such as a rectangular toolpath bounding the outside of a box, or an inner perimeter, such as a circular toolpath surrounding a hole. Some slicing programs allow for multiple perimeter paths to be printed alongside one another, such that the entire region of perimeter paths is referred to as the perimeter. Fig. 6.18 shows a layer with outer and inner perimeter beads.

Inset closed-loop paths reside inside of the perimeters, either offset towards (for outer perimeters) or away from (for inner perimeters) the centroid of the polygon. The term "inset" arose as shorthand for "inward offset," because insets move towards the inside of the layer polygon. The inset region typically contains multiple paths to help add strength and rigidity to the object. Insets and perimeters can sometimes be called "vertical shells" because they create solid shells around the sides of the object. Fig. 6.19 shows inset beads added to the geometry of Fig. 6.18.

Perimeters and insets are both found using the same process for closed-loop paths: the erosion and dilation method mentioned earlier. The reason for having two different paths of seemingly identical geometry and pathing is to allow the printing process to apply different settings to each region. With perimeter being the outer most bead on a

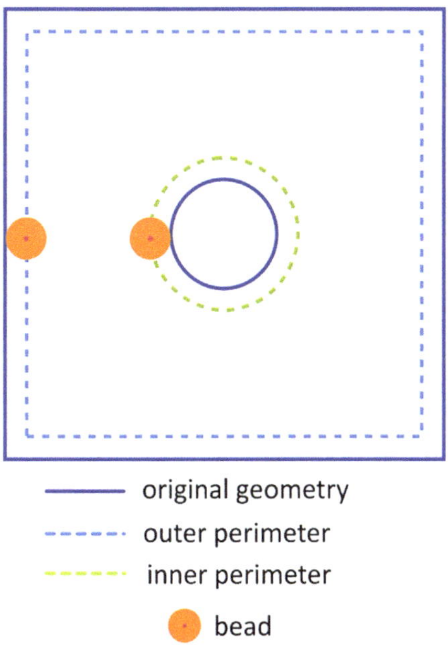

FIG. 6.18 A square with inscribed circle and the resultant pathing when adding a perimeter.

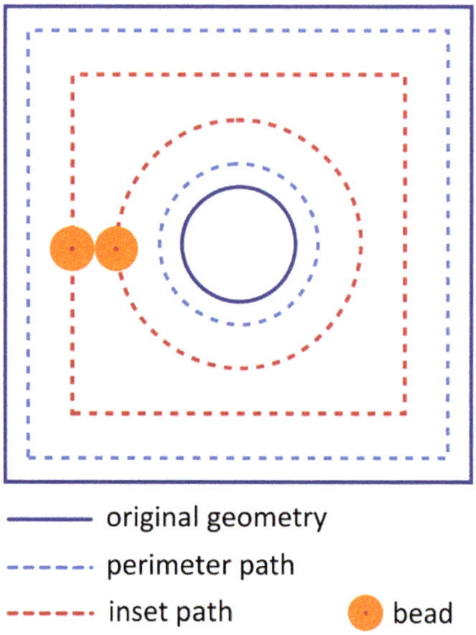

FIG. 6.19 When a single inset is added to Fig. 6.18, one closed-loop inset path is added offset from each boundary of the layer, creating two inset paths.

layer, it is often the only exposed portion of the object. Therefore, it is desirable to print this outermost bead, the perimeter bead, at a slower speed and extrusion rate so that the bead geometry is more consistent and presents a better surface finish. However, it's not desirable to use this slower speed for every closed-loop contour path. That is where having a second region, or second definition of a closed-loop toolpath (the inset), becomes valuable. The process can be configured such that insets use a larger bead width, faster speed, and higher extrusion rate so that more area is covered in less time. Alternatively, it is sometimes beneficial to print a larger bead width for the perimeter than for the inset, such that the perimeter is "overbuilt," and the extra material can be machined back to the desired size and tolerance. The design of pathing for machining and hybrid operations will be discussed further in *Chapter 17: Pathing considerations for hybrid systems*.

Conclusion

Closed-loop paths have start and end points that share coordinates and are the first path type to be generated when fitting toolpaths to a layer. To find the location for a closed-loop path, the centerline is found through an erosion-and-dilation method that is applied to the bounding edges of the layer polygon. The nature of this offsetting process requires that the object's geometry be at least two-bead-widths thick for a closed-loop path to fit within the bounds of the polygon. Many layers will contain multiple closed-loop paths, which can be divided into perimeters and insets so that optimal settings can be applied to the exposed, outermost portions of an object (perimeters) and to the more interior portions (insets). Any region of a part that violates the two-bead-width constraint cannot be fit with a closed-loop path, but it may be possible to fit an open-loop path to the region. The next chapter will discuss open-loop paths and how they are generated.

References

[1] D. Zhao, W. Guo, Shape and performance controlled advanced design for additive manufacturing: a review of slicing and path planning, J. Manuf. Sci. Eng. 142 (1) (2020) 010801.

[2] G. Gomez, et al., Generation of continuous hybrid zig-zag and contour paths for 3D printing, Int. J. Adv. Manuf. Technol. 119 (11) (2022) 7025–7040.

[3] J. Hu, Study on STL-based slicing process for 3D printing, in: 2017 International Solid Freeform Fabrication Symposium, University of Texas at Austin, 2017.

[4] A. Messing, et al., Using skeletons for void filling in large-scale additive manufacturing, in: 2017 International Solid Freeform Fabrication Symposium, University of Texas at Austin, 2017.

[5] M. Bertoldi, et al., Domain decomposition and space filling curves in toolpath planning and generation, in: 1998 International Solid Freeform Fabrication Symposium, 1998.

CHAPTER

7

Space-filling toolpath generation

Alex Roschli[a], Michael Borish[a], Abby Barnes[a], Peter Wang[a], Thomas Feldhausen[a], and Eric MacDonald[a,b]

[a]Manufacturing Demonstration Facility, Oak Ridge National Laboratory, Knoxville, TN, United States, [b]University of Texas at El Paso, El Paso, TX, United States

Abstract

Space-filling toolpaths are used to fill whatever space remains on a layer after closed-loop contours have been planned. They provide structure to the object, both supporting the geometry and creating solid surfaces. As such, there are two main types of space-filling toolpaths: infill and skin. Infill paths are sparse and meant to cover a large area quickly. Infill is typically not visible once a print is complete, because infill is covered by skins. Skins are the solid space-filling toolpaths meant to solidify the top and bottom of the object, which, unlike the sides of the object, are not completely covered by layering contours. Space-filling paths are typically planned by projecting a pattern over the layer, clipping the pattern at the boundary of the object, and then linking the remaining portions of the pattern. A subcategory of skin paths, called gradual infill, can be employed to densify the infill when approaching a top skin layer so that the print path of the top skin is sufficiently supported. Space-filling paths come in a variety of patterns to optimize how the space is filled. This chapter will discuss space-filling path categories, how to find the space for each path type, and how to apply the path type to generate toolpaths.

Introduction

To ensure the definition of the perimeter of a layer, the tool pathing for the layer typically begins with closed-loop contours. After the contours have been found, the interior of the layer polygon is still left unplanned. This center, or core, of each layer can often be quite a large area. To optimally fill it, space-filling toolpaths are used. Space-filling paths come in many varieties and patterns and can be tailored to fit whatever polygon shape is left from other pathing [1]. Like closed-loop paths, space-filling paths have modifiers that can be applied after the pathing trajectories

https://doi.org/10.1016/B978-0-443-15286-3.00009-6

are generated. These modifiers are discussed in *Chapter 10: Path modifiers*. Likewise, the methods described in *Chapter 11: Travels, optimizations, and ordering* can also be applied to space-filling toolpaths.

Types of space-filling toolpaths

Space-filling toolpaths come in two major categories: infill and skins. Both infill and skins can be generated using various space-filling algorithms. The main differences between the two are the density and location/implementation of the toolpaths. Infill pathing is typically sparse, meaning that the toolpaths do not cover 100% of the area they are meant to fill, but the user can define the infill as fully dense. The void space produced from sparse infill pathing is intentional, to save print time and material while still providing a fill that will reinforce the object. Skin pathing is 100% dense, or fully solid, with no spacing between beads, such that the entire area is filled with toolpaths.

Infill toolpaths are typically used within the center, or core, of the printed object and are not visible when the object is finished. This is because skins are printed on top (up skin or top skin) and below (down skin or bottom skin) infill layers so that the infill is fully encapsulated. As such, skins (also known as "horizontal shells" because they form solid shells on the horizontal plane) are typically used for the outermost layers and infill is used for the remaining layers.

The term "skin count" is used to describe how many layers of skin will be produced for each top or bottom surface. A top skin count of three implies that the topmost layer *and* the two layers below it will all be skins. This is not limited to the last three layers of an object; it includes the three topmost layers of any upward-facing surface. Likewise, bottom skins do not occur only on the beginning layers of a part. Bottom skins occur on every downward-facing surface. Fig. 7.1 shows top and bottom skins occurring at different heights within a single object. The following section will talk about finding skin locations within a layer.

Finding the space to fill

After all closed-loop paths have been generated, a remaining layer polygon is calculated to represent the area that must be filled with space-filling toolpaths. Before toolpaths can be fit to this polygon, the region, or regions, of the polygon must be classified as skin or infill. This means that the polygon needs to be assigned a type of pathing: either skin, infill, or a combination of the two wherein the polygon would be divided

FIG. 7.1 An object with both top and bottom skins occurring on multiple layers.

into multiple, smaller polygons that are each assigned skin or infill pathing.

When looking at the polygon for space-filling pathing, the first step is to determine if any area of the polygon should be assigned a skin, because skins take priority over infill. Infill is the lowest priority because it is always used as a "filler" and is assigned to whatever area is leftover in a layer after all other paths are planned. To determine if part or all of the remaining layer polygon should be a skin, the algorithm must have some information about the entire build, including previous and future layer geometry, and the number of desired top and bottom skins, as defined by the user in the slice settings. If none of the polygon should be a skin, the entire polygon can become infill.

Starting with top skins, the algorithm looks at the current layer's remaining geometry (the geometry without any planned paths) as well as each of the layer polygons of the future *N* layers, where *N* is the desired number of top skins. These future layers don't have to have paths planned yet, and don't have to be assigned regions such as skin or infill. The current remaining layer polygon is compared each future layer polygon using a Boolean polygon operation called a difference, which subtracts the area of the future layer polygon from the area of the current layer polygon and checks for any remaining geometry. If the entire area of the current layer exists on all of the future layers, then the current layer does not need a skin and can be planned with infill. If the entire area does not exist on one or more of the future layers, then the entire area of the current layer will be made into a skin.

A third scenario occurs if part of the current layer area does not exist on future layers but part of the area does exist. All areas that don't exist on future layers are made into a polygon, or polygons, and designated as top skins. All areas that continue to exist on future layers are made into a polygon, or polygons, and designated as the remaining layer polygon.

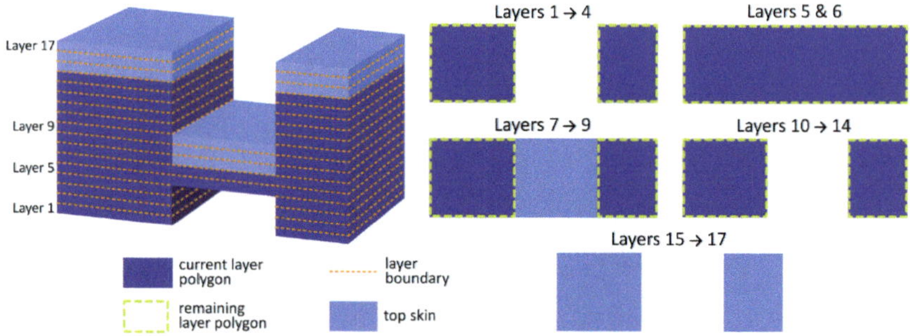

FIG. 7.2 Top skin generation for an object with varying top levels.

This remaining layer polygon is to be used in the next step, the generation of bottom skins. Fig. 7.2 shows all of the layers of an example object during the top skin generation process where the top skin count is 3.

The process for bottom skins works the same way as for top skins, but instead the comparison is with the previous N layers where N is desired number of bottom skins. In this comparison, the current remaining layer polygon does not include areas designated as top skins, but for illustration purposes, Fig. 7.3 shows the downskin generation process with top skins disabled, for the same object as is in Fig. 7.2.

After bottom skin generation is complete, all the remaining area of each layer polygon can be designated as infill, or an additional step can be included: gradual infill. Gradual infill can be viewed as a subset of skin and is used as a transition region between sparse infill and solid skins. Processes with low-viscosity extrusion struggle to extrude a solid skin layer on top of a very sparse infill layer because of all the void space that the skin pathing can fall into. Gradual infill allows for a gradual densifying of the infill when approaching a top skin layer, so that the top skin has a higher chance of succeeding. The density and number of infill steps is user-defined such that the infill gradually becomes denser as it approaches a top skin.

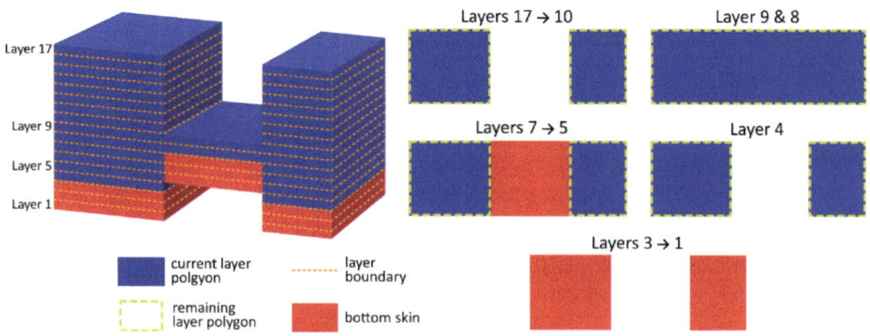

FIG. 7.3 Bottom skin generation process for an object with varying bottom levels.

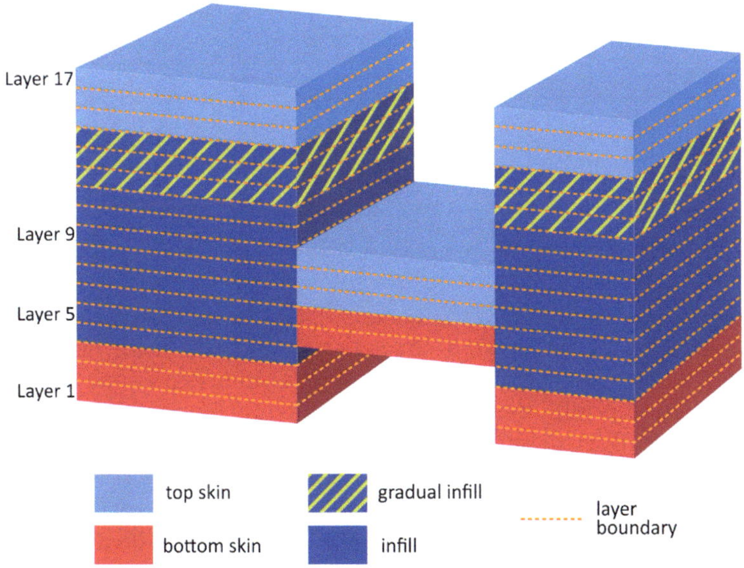

FIG. 7.4 Generating top and bottom skins for an object with gradual infill enabled.

Gradual infill is planned after bottom skins and before infill. To find areas of gradual infill, the same process for finding top skins is used. The number of steps of gradual infill plus the number of top skins becomes N, the number of future layers to look at for comparisons. The comparison isn't looking for future layers with skins; it is looking for future layers where the current layer geometry doesn't exist, just like how top skin generation works. As with bottom skin generation, existing skin regions in current layer polygons are not included as part of the remaining layer polygon when used for comparisons because gradual infill cannot overwrite an existing (top or bottom) skin region. Fig. 7.4 shows the same object from Figs. 7.2 and 7.3 with all three processes (three top skins, three bottoms skins, and three gradual infill) applied in sequence. Note that the bridge now only has two bottoms skins, because the top skin of Layer 7 is not counted in the layer's remaining polygon when calculating for bottom skins. After gradual infill regions have been found, the remaining solid purple areas of the object are finally designated as infill.

Space-filling toolpath settings

Infill pathing is influenced by a variety of settings that are defined by the user before initiating the slicing process, the most important of which is the infill density, or the infill line spacing. Infill density is used to define how dense the infill pathing should be. It is represented as a percentage,

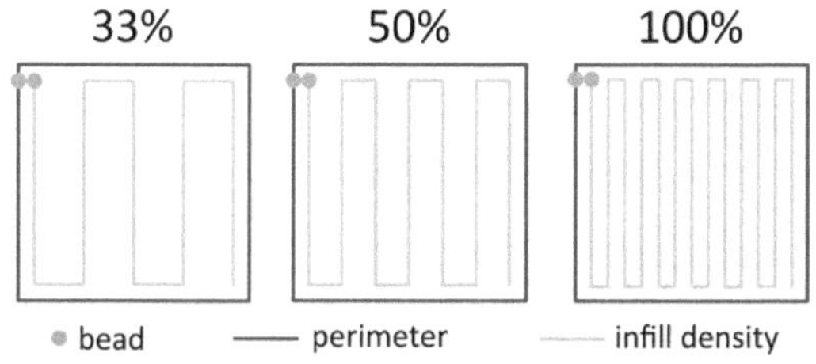

FIG. 7.5 A square profile with various infill densities.

such as 50%, to define the amount of area to be covered by infill paths and, inversely, the amount of void space between beads. Infill line spacing defines the same thing, but gives the user the ability to define a physical distance between infill beads. A line spacing of 0.5 with a bead width of 0.25 would result in an infill density of 50%, whereas a line spacing of 0.25 with the same 0.25 bead width would generate fully dense infill, 100%. Fig. 7.5 shows a variety of line spacing and density values. Gradual infill also has a density value, but it is auto-calculated based on the infill density and the number of steps between the infill and the skin. For example, if the infill density is 50% and the number of steps is 3, the three layers of gradual infill would have the following densities: 62.5%, 75%, and 87.5%.

Bead width is another important setting for fill pathing. As previously mentioned, this is used to calculate the covered area by determining how much space is taken up by each path segment. Overlap is another setting which can be used to enlarge the polygon of the fill region so that the pathing slightly overlaps the bounding regions. This is helpful for ensuring the fill pathing has a good bond with surrounding pathing and is not easily broken away. Additional settings exist for defining the fill pathing, but all are specific to the specific fill pattern being used. Fill pattern, another important setting, is defined in the following section, which discusses various patterns and how they can be applied to each type of space-filling toolpath.

Fill patterns

For each fill pathing region, a variety of patterns can be applied. There are less options available for skin patterns than for infill and gradual infill because skin patterns must be 100% dense (fully solid). Due to the nature of how space-filling toolpaths can be generated, there is an unlimited number of patterns and strategies to fill the space. The following subsections discuss various commonly used patterns and how they can be applied to infill, gradual infill, and skin polygons. The list of patterns is not exhaustive but is meant to represent the most popular strategies and give a complete view of many different possibilities.

Rectilinear

Rectilinear fill is the most popular infill pattern because of its simplicity and effectiveness. The fill pattern, similar to that seen in Fig. 7.5, is a series of lines resembling a raster pattern to fill the interior of the object. Rectilinear fill can be sparse or solid, so it can be used for all three types: infill, skin, and gradual infill.

To generate rectilinear fill, a group of equal-length parallel lines is projected over an area that is defined by the minimum and maximum boundaries of the entire part. Then the lines are clipped by the polygon to be filled for the specific layer. The lines are spaced according to the line spacing, or the spacing is calculated based on the commanded density. Finally, the lines are connected so that one continuous path is drawn from one end to the other. Fig. 7.6 shows the entire process of generating rectilinear infill.

This approach generates layers of infill that are specific to the specific part being constructed, meaning that the pathing will generate identically for each instance of an object. Ten equivalent cubes can be sliced at once and they will all have identical infill pathing. This is the most desired approach, but certain situations require the ability to shift the fill pathing within the object. To add this ability, the fill pattern is generated across the entire build area of the configured 3D printer, then clipped to the bounds of the layer's fill polygon. This means that moving the object within the workspace generates a different location for the fill paths, based on which paths intersect the object. Fig. 7.7 shows the process of generating infill based on the part versus generating based on the printer area.

If the pathing cannot all be connected, such as the object shown in Fig. 7.8A, a travel move can be added. Travels are paths during which no extrusion occurs and are discussed more extensively in

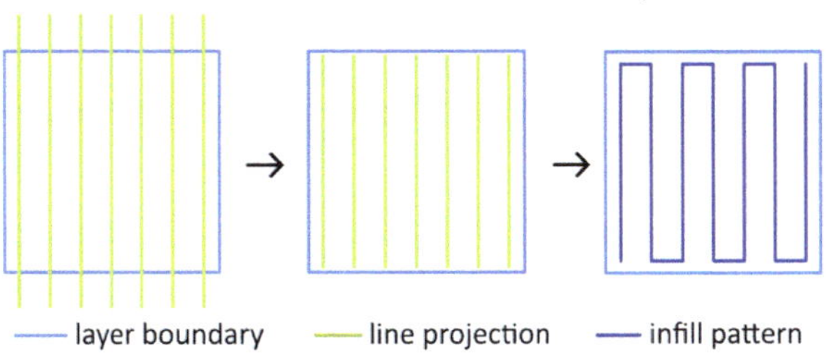

FIG. 7.6 The process of generating infill: projecting the fill pattern, clipping to the layer boundary, and connecting the toolpath lines together.

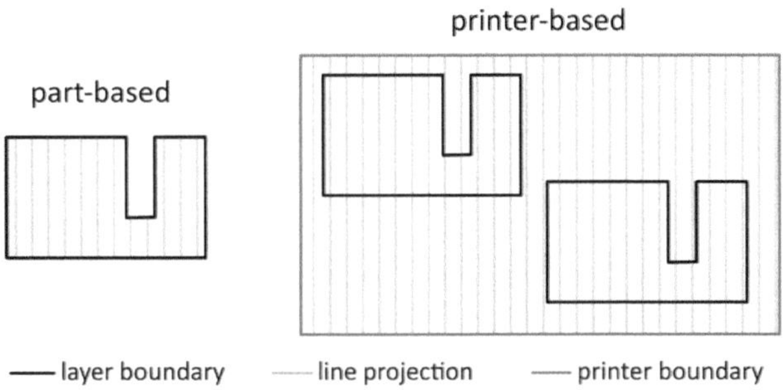

FIG. 7.7 Generating infill over just the part and generating infill over the entire printer area.

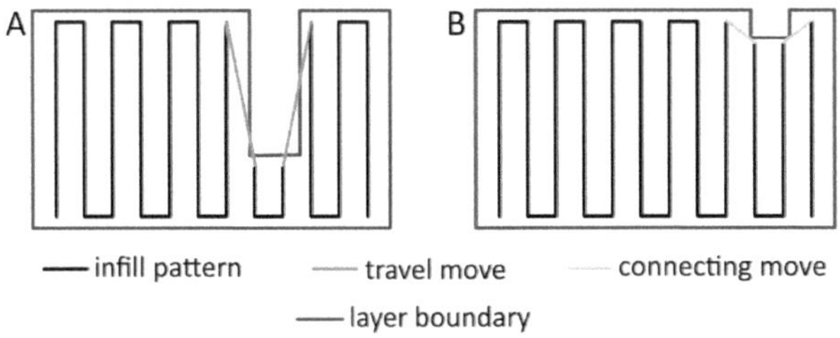

FIG. 7.8 Travel moves (A) and infill moves (B) added to infill to optimize the pathing based on the move distance.

Chapter 11: Travels, optimizations, and ordering. If the two separate paths are close enough together, a connecting extrusion move can be added to prevent a travel move. This connecting move will only be an extrusion move if the entire area covered by the connecting move is contained within the layer bounds. A connecting move that goes outside the bounds would cause extrusion in areas that aren't meant to be printed, which is not desirable. An example of a connecting move can be seen in Fig. 7.8B.

Fig. 7.8 shows infill lines going top to bottom, or at a 90-degree angle. The pattern can easily be rotated to fill at any angle as desired by the user. Fig. 7.9 shows infill in a square at a 45-degree angle.

The infill angle can also be rotated from one layer to the next, which helps improve the fill of the object by applying material at different angles. Rotating by 90-degrees helps to make objects more uniform in the X and Y planes, particularly with materials that contain fiber. During the printing process, the fibers naturally become aligned with the extrusion direction, so the rotation helps distribute the strength of the fiber between the X and Y directions. This 90-degree infill rotation from one layer to the next is so common that it is considered the standard implementation for

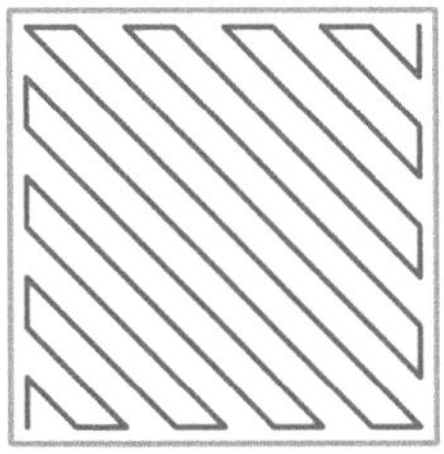

FIG. 7.9 A square object with 45 degree infill.

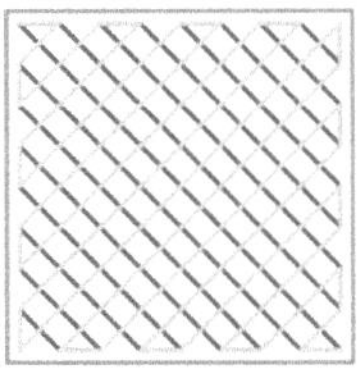

FIG. 7.10 Two layers of infill rotated by 90 degrees.

rectilinear infill. Fig. 7.10 shows the shape from Fig. 7.9 with another layer of infill added on top, after being rotated 90-degrees.

Grid infill

Grid infill can be considered a subset of the rectilinear fill pattern because it uses the rectilinear fill algorithm to create a square grid infill pattern. Grid infill is generated by applying the rectilinear infill pattern at an angle (0-degree, for example), then applying the rectilinear fill again on the same layer but rotated by 90-degrees to create squares. For each rectilinear pass, the infill density or line spacing is cut in half. This is because there are two sets of rectilinear paths per layer, and they add together to meet the total infill density. Fig. 7.11 shows an implementation of grid infill.

A downside to grid infill is that the extrusion is constantly crossing over itself. For certain processes, such as metal directed energy deposition (DED), this isn't a problem because the DED process can remelt the material when crossing over. However, polymer processes cannot remelt the

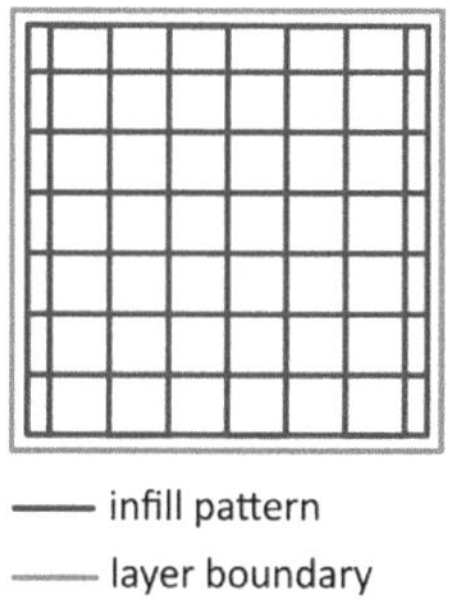

FIG. 7.11 The grid fill pattern.

previous layer when crossing it. If the layer is large, the first pass of rectilinear fill will have cooled, and the bead will break each time it crosses the previously printed material. For smaller layers, the first pass may remain hot and flexible enough to not break the bead, making the bead more likely to bond.

Grid fill cannot effectively be printed at high density values, and cannot achieve 100% density. Therefore, the grid pattern is only used for infill and gradual infill; it is not available for skin paths.

Hexagon-and-triangle fill

Hexagon-and-triangle is another subset of the rectilinear pattern, but it involves three passes of rectilinear fill that are rotated by 60-degrees each. The result is a repeating pattern that creates hexagon and triangle shapes throughout the fill. Like grid fill, the hexagon-and-triangle fill cannot be made fully dense and cannot be used for skins. The line spacing for each of the three passes is one third of the input line spacing from the slicing settings. Fig. 7.12 shows an implementation of the hexagon and triangle fill.

Triangle fill

Triangle fill works just like the hexagon-and-triangle fill, but with one modification. The third set of rectilinear paths is shifted by half of the line spacing distance. This allows the third set of lines to pass through the same node where the first and second set of lines intersect. As mentioned with the grid, this intersection of multiple lines does not work well for extrusion systems and is likely to break the bead each time it crosses a previously printed bead. Fig. 7.13 shows the triangle fill pattern.

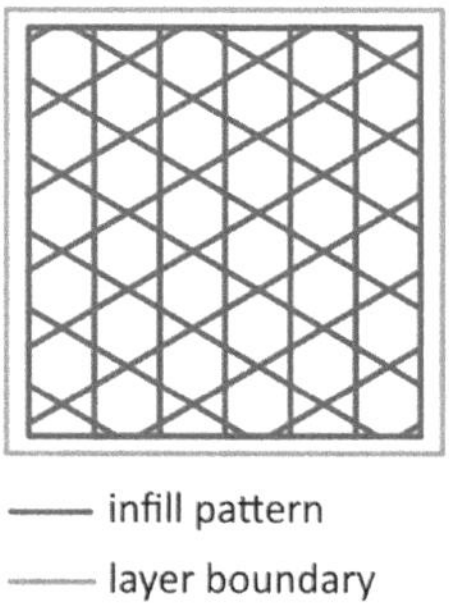

FIG. 7.12 The hexagon and triangle fill pattern.

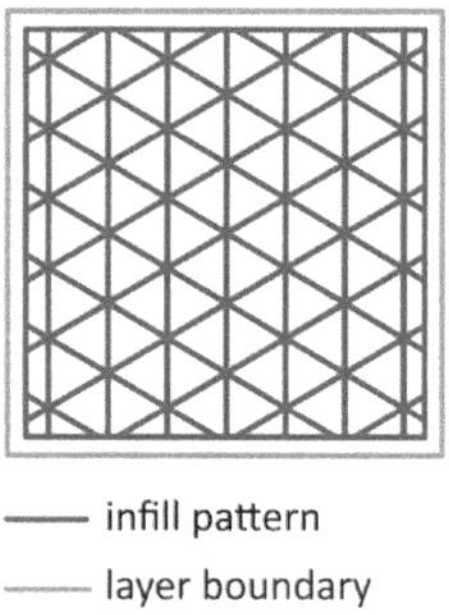

FIG. 7.13 The triangle fill pattern.

Concentric

Concentric fill uses closed-loop fill paths in which each successive path is contained inside the previous path. This fill is generated using the same process used to find closed-loop contours, which was discussed in *Chapter 6: Closed-loop toolpath generation*. This concentric pattern can be used to generate a fully solid layer for skins and is a popular pattern because of the smooth surface finish it creates (Fig. 7.14A). The pattern can also be used for sparse infill by making the erosion offset distance equal to the line spacing (Fig. 7.14B). However, concentric fill is rarely used for infill because each pass of the infill generates a path that stands alone and doesn't connect to other pathing. These individual paths do not function well to support the entire structure because they are disconnected. Concentric fill also does not work well for gradual infill because the changing fill density would prevent the layers from lining up, meaning the fill may not have a previous layer to rest on. The concentric path ordering can be inside-out or outside-in, depending on the object and/or user preferences.

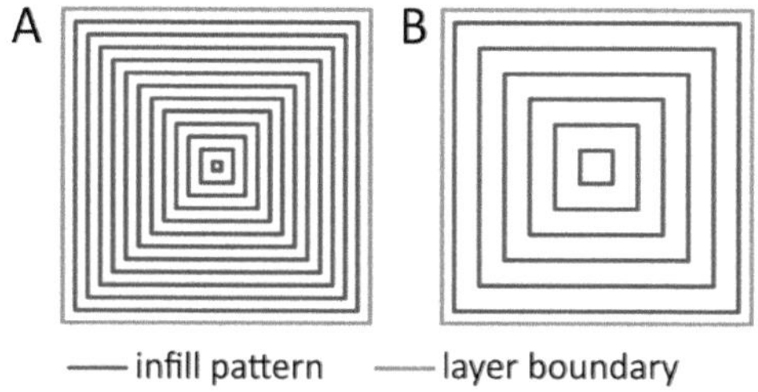

FIG. 7.14 The concentric fill pattern with high (A) and low (B) density.

Honeycomb

Honeycomb is a popular fill pattern inspired by and based on the natural honeycomb pattern: repeated hexagons. Like the rectilinear fill, the pathing is found by projecting the pattern across the bounds of the object and then clipping it at the boundaries of the layer's fill polygon. Rather than projecting lines, the projection uses two halves of the pattern, one vertically flipped from the other, shown in Fig. 7.15A. When these patterns come together to make infill, the result has two adjacent beads for all of the 0-degree segments of the path and just a single bead for the remainder of the path segments (Fig. 7.15B). For an object with the same cross-section for each layer (such as a cube), the infill pattern is identical layer after layer so that the infill paths sit atop one another. The infill density defines the size of the infill hexagon such that an appropriate amount of void space is created. The honeycomb pattern can be modified to form a 3D shape over the progression of many layers. This is done by slightly altering the

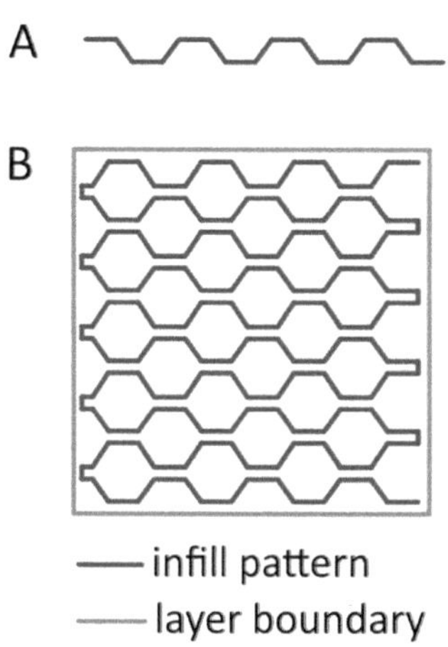

FIG. 7.15 (A) One half of the honeycomb fill pattern; (B) the resulting infill pattern when the half honeycomb pattern is projected with alternating orientations.

hexagon size from one layer to the next so that the various hexagons are all either opening (getting larger) or closing (getting smaller) from one layer to the next. Honeycomb fill cannot be made fully dense and therefore cannot be used for skin pathing. Gradual infill does not work for the honeycomb pattern because densifying the infill makes the hexagons smaller and typically results in infill pathing that is not printed directly over the previous layer, leaving nothing to support the paths.

Gyroid

The gyroid surface was originally developed by Alan Schoen in 1970 [2] and is described as the only triply periodic minimal surface (TPMS) that does not contain straight lines or plane lines of curvature. The gyroid pattern can be printed intersection-free, meaning that the toolpaths on each layer never cross and break the bead [3]. The wavy shape of each infill pass is governed by Eq. (7.1). The number of passes and the distance between passes is governed by the fill density or line spacing.

$$\sin x \cos y + \sin y \cos z + \sin z \cos x = 0 \tag{7.1}$$

The gyroid pattern is implemented similar to the rectilinear fill pattern, with a 90-degree rotation from one layer to the next, but the lines are represented as waves and the position shifts slightly from layer to layer. The result is a 3D pattern that creates nearly equivalent mechanical properties in all directions. Fig. 7.16A shows one layer of the wavy gyroid infill pattern. The following layer will follow the same pattern but is rotated 90 degrees (Fig. 7.16B). A third layer will be identical to the first, but the pattern will be shifted along the Y-axis. This shift is what allows the gyroid surface to be created. The shift cannot be larger than one half of the bead width because the bead being deposited must have a bead below it to rest on top of.

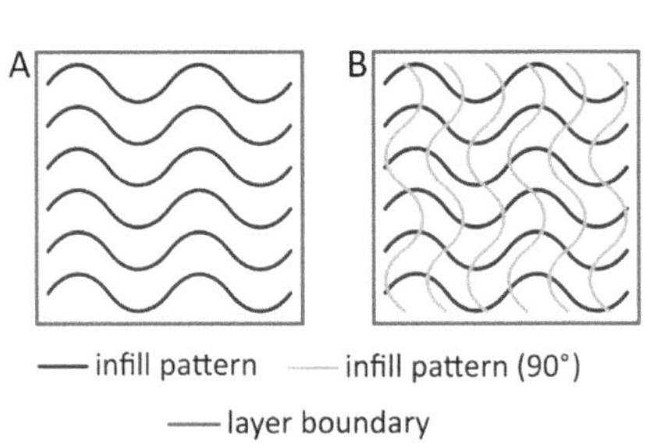

FIG. 7.16 (A) One layer of the gyroid fill pattern; (B) the next layer of the gyroid fill pattern, which has been rotated 90 degrees.

Due to the nature of how the gyroid pattern is generated, it only makes sense to use the pattern for sparse infill. The pattern would not work for a fully dense fill, such as a skin. The pattern also doesn't work well for gradual infill because gradual infill changes the infill line spacing (or infill density) from layer to layer. The paths would likely print successfully, but they wouldn't be well aligned with the paths from the previous layer, which would disrupt the gyroid surface.

Hilbert curve

Up to this point, most of the fill patterns have had symmetry about the *X* and *Y* axes. They are repeatable and create uniform mechanical properties within the plane of the 2D layer. Hilbert curve infill is different in that it is not a uniform pattern across the whole layer, but it can have symmetry across the *Y* axis. The Hilbert curve pathing is identical from layer to layer for an object with a consistent cross-section (such as a cube). The Hilbert curve is a continuous, fractal, space-filling curve defined by David Hilbert in 1891 [4]. The Hilbert curve makes a series of "cup" shapes. The number of cups is based upon the order of the curve, with a first order curve having a single "cup" shape. The center of each higher order cup aligns with the points of the cup(s) from the previous order of curve. This means that each order increases the number of cups by a magnitude of four, because each cup has four points. A higher density infill requires a higher order Hilbert curve. Fig. 7.17A shows a first order Hilbert curve with a single cup. Fig. 7.17B shows a second order with four "cups" and three joining segments that are also print segments, to make one continuous infill path. Fig. 7.17C shows a third order Hilbert curve, which has 16 cups with 15 joining segments.

The Hilbert curve fill creates large continuous voids within the infill that can then be backfilled with a low viscosity material for increased strength. Because of how this pathing works, it cannot be used for solid fill skins. A Hilbert curve does also not work well for gradual infill. Like

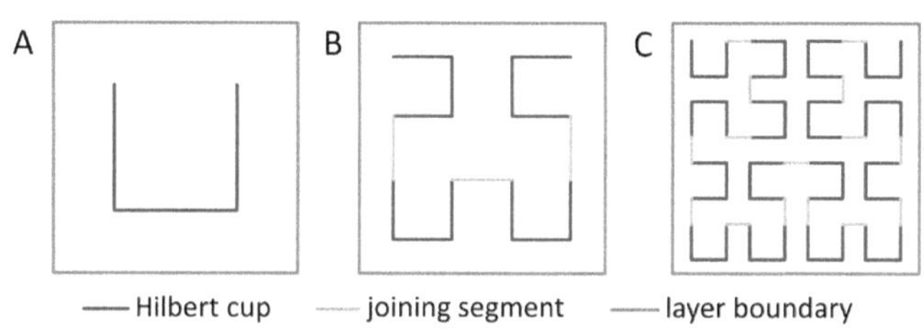

FIG. 7.17 One (A), two (B), and three (C) orders of the Hilbert curve.

with other patterns, the gradual densifying of the pattern causes the path location to shift, which prevents the pathing from having a previous layer as support.

Conclusion

Space-filling toolpaths are used to fill the interior of the layer and to provide structure and rigidity to the object. There are three types of space-filling toolpaths: infill, gradual infill, and skins. All three types can exist within the same layer, depending on the object geometry and user settings. Skins are fully dense, space-filling paths that are typically implemented with either a rectilinear pattern or concentric pattern. Infill and gradual infill have sparse density that is user-defined, and can use a wider variety of patterns, such as grid, honeycomb, and gyroid. Space-filling toolpaths are the second category of paths that must be planned for construction of an object. Other toolpaths, such as open loop paths, may be necessary to ensure a successful print, which will be discussed in the following chapters.

References

[1] M. Bertoldi, et al., Domain decomposition and space filling curves in toolpath planning and generation, in: 1998 International Solid Freeform Fabrication Symposium, 1998.
[2] Schoen, A.H. Infinite periodic minimal surfaces without self-intersections. No. C-98. 1970.
[3] K. Große-Brauckmann, W. Meinhard, The gyroid is embedded and has constant mean curvature companions, Cal. Var. Partial Differ. Eq. 4 (6) (1996) 499–523.
[4] D. Hilbert, On the continuous mapping of a line onto a patch, in: Third Volume: Analysis Basics of Mathematics Physics Miscellaneous, Springer, Berlin, Heidelberg, 1935, pp. 1–2.

CHAPTER

8

Open-loop toolpath generation

Michael Borish[a], *Alex Roschli*[a], *Abby Barnes*[a], *Eric MacDonald*[a,b], *Peter Wang*[a], *and Thomas Feldhausen*[a]

[a]**Manufacturing Demonstration Facility, Oak Ridge National Laboratory, Knoxville, TN, United States,** [b]**University of Texas at El Paso, El Paso, TX, United States**

Abstract

Open-loop paths, as opposed to closed-loop paths, have start and end points that are at different locations. Open-loop paths, commonly called "skeletons," can be generated through multiple approaches. Most approaches utilize a shape's medial axis transform and build appropriate pathing around this representation. In this chapter, a specific implementation consisting of three major steps will be discussed. These steps are Voronoi diagram generation, skeleton pruning, and path consolidation. With the major components of the implementation covered, two additional optimizations will also be described. These two optimizations, chamfering and smoothing, will be highlighted as ways to improve overall pathing quality.

Introduction

With both closed-contours and space-filling toolpaths generated, path planning proceeds to the generation of open-loop paths. Open-loop paths are generally referred to as "skeletons" and, like every geometric processing step in this book, there are multiple ways to tackle their generation. In general, there are two types of skeletons encountered in AM: straight and medial axis. Both skeletons are types of topological skeletons, which are thin versions of a shape that are equidistant to the shape's boundaries. Topological skeletons can emphasize various properties, such as length, width, connectivity, and topology (obviously), depending on construction technique.

Straight skeletons are composed of straight-line segments, hence the name, and are generated by continuously shrinking the shape until it

https://doi.org/10.1016/B978-0-443-15286-3.00024-2

collapses into a skeleton-like structure. The straight skeleton describes the change in shape and the paths taken by the vertices as they move inward. Alternatively, the medial axis can include curves and is generated via the medial axis transform (MAT). The medial axis represents the points that are equidistant to the boundaries of the shape and can be thought of as a subset of the straight skeleton. Because these two types of skeletons are highly related, it is possible to convert between them. That is why this chapter will focus on the medial axis. A full discussion of the medial axis can be found in the work of Blum who originated the concept for shape descriptors [1]. At a high level, the medial axis for a polygon is the set of points that are each the center of a distinct circle. The circles are defined such that each is contained inside the polygon and is tangent to the polygon at a minimum of two points. This produces a characteristic "skeleton" inside the polygon, like what is shown in Fig. 8.1. The set of points and their associated circles is collectively called the medial axis transform (MAT), which is a complete representation of the original shape and can be singularly used to reconstruct the shape.

To generate the medial axis for use in skeletonization, there are three general approaches: distance transformation, thinning, and Voronoi overlay. In a distance transformation, a distance metric is used to calculate the amount of offset of every point within a shape from the boundaries of the shape, and the medial axis is composed of the points that are maximally offset from the shape's boundaries. The classic metaphor is to consider everything in an image or shape to be made of grass or trees. If you set fire to every point along the outer contour simultaneously and let it burn, the medial axis would be the line where the fire meets itself. Thinning, on the other hand, is an iterative reduction that is based on the contours of the shape in question. Some algorithms of this type are like the offsetting (erosion) process described in the previous chapter. The last option, a Voronoi

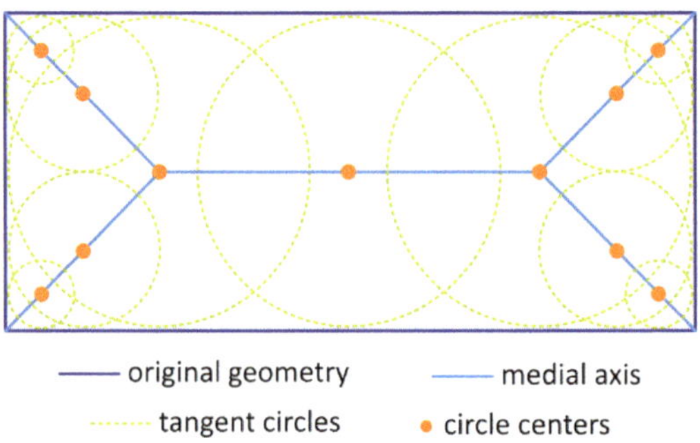

FIG. 8.1 Medial axis transform (MAT) was generated using the overlay generated from the tangent circles.

overlay, creates a Voronoi diagram (VD) of the polygon, and the medial axis is extracted from this diagram. In fact, the medial axis is a subset of the Voronoi diagram. It is this Voronoi overlay approach to skeletal pathing that will be described in more detail later in this chapter.

Geometry identification

Before beginning to lay out open-loop paths, the portions of the geometry that need to be fit with open-loop paths must be found. Recall that, during the erosion-dilation process that is used to generate closed-loop contours, it is possible to lose geometry, like the example shown in Fig. 8.2, which is the same as Fig. 7.4. The negative area highlighted by a dashed line in the rectangular portion of the geometry is unsuitable for closed-loop paths because it does not meet the requirements of the two-bead-width constraint and is self-intersecting. It would be possible to fill such an area with infill, but this would require careful selection, placement, and rotation of an infill pattern. For example, a line pattern at a 90-degree rotation could be used, but because infill patterns are static, meaning the lines of the pattern cannot be altered individually, it would be difficult to get the pattern to completely fill the space. This is because it is unlikely that the lines of the pattern will perfectly match up with the lines of the geometry. Therefore, for cases such as this, a skeleton is more appropriate because it is a dynamic path that can be more readily fit to the geometry.

The same remaining layer polygons that resulted from the previous path-generating steps, discussed in the previous chapters, are used to track the geometry that needs to be filled with skeletons. After closed-loop and space-filling paths have been planned for a layer, the layer is again checked to see if there is any geometry in the remaining layer polygon. Like the example in Fig. 8.2, there are usually portions of the polygon that

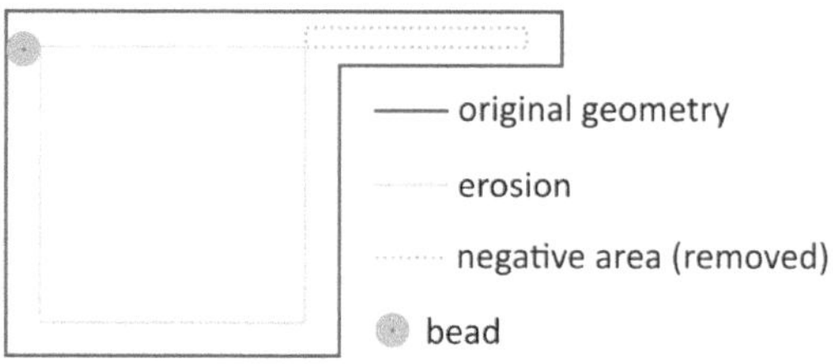

FIG. 8.2 Polygon offset with closed contours. The negative area in the top-right will be removed leaving a rectangle to be filled with closed contours. This small rectangular area can be filled with a skeleton.

were unsuitable for the previous two path types because of the geometry and/or various infill settings, such as the infill pattern. These areas therefore remain unpathed. It is these remaining portions that need to be evaluated for the addition of open-loop paths.

One important note about this remaining geometry: It is often the case that numerical inaccuracies arise due to the number of geometric manipulations that take place in the repeated applications of erosions and dilations. As a result, the remaining layer polygon is very likely to include many small polygons. Some of these polygons may be so small that no true geometry was lost, yet these polygons must still be dealt with. To address this issue, the remaining layer polygon should be simplified in much the same way operations like smoothing are applied to cross-sectioning.

One straightforward simplification is to evaluate the width of each portion of the remaining layer polygon against a user-specified threshold. This threshold can reasonably be set to some percentage of the bead width. If a portion of the remaining layer polygon does not meet the threshold, then the quality of the print is likely improved by discarding that portion. For example, if a portion of the polygon is only 30% of the bead width wide, then it is unlikely it could reasonably be filled with the construction process, and it can safely be ignored.

With this final simplification in place, skeleton generation can proceed. Basic skeleton generation can be accomplished with three steps: Voronoi diagram (VD) generation, skeleton pruning, and path consolidation. For a formal and detailed explanation of VD's, Aurenhammer provides an excellent survey [2].

Voronoi diagram generation

At the most basic level, a VD is a partition of a plane into regions. Each region of a VD surrounds a point on the plane and defines the area of the plane that is closer to that point than any other point on the plane. Mathematically, for a finite set of points $P=\{p_1,p_2,\ldots,p_n\}$ in a Euclidean plane, a Voronoi diagram is a partitioning of the plane into R_k regions such that each region consists of every point whose distance to p_k is less than its distance to any of the other points in P. An example of this plane is shown in Fig. 8.3.

VD generation is well understood and supported by multiple software libraries. Two of the most well-known are Boost and Computational Geometry Algorithms Library (CGAL) [3,4]. Readers interested in a deep dive into the implementation of VDs can consult the documentation for either library. In general, though, VDs are computed one of three ways: incremental, divide-and-conquer, or sweep line.

The incremental approach for constructing a Voronoi diagram involves the incremental introduction of one point, from a set of points, onto a

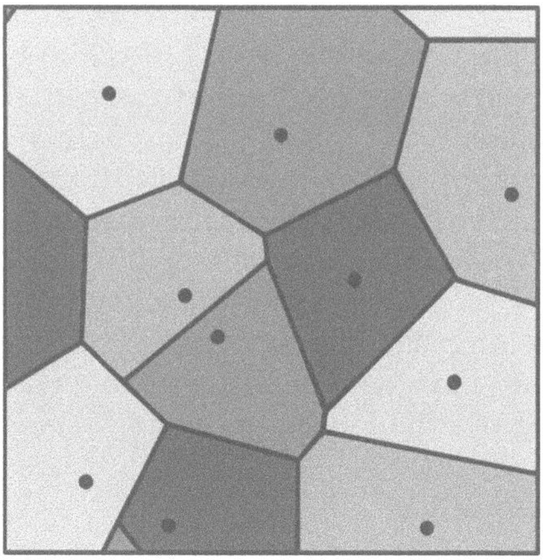

FIG. 8.3 Voronoi diagram (VD) of a plane. Rather than using a full plane, we will develop a VD overlay using the polygons in our slicing process.

plane. It defines the VD region around each point before introducing the next point, adjusting the existing regions as necessary. For example, starting with a point p_1 on a plane, a new point p_2 is introduced onto the plane. A Voronoi edge is created that bisects p_1 and p_2, forming a boundary between their respective regions. At this point, the VD is just the plane divided into two regions. Then, another point p_3 is introduced and the pre-existing points (p_1 and p_2) are evaluated to identify which is closest to p_3. A Voronoi edge that bisects p_3 and the closest point is created. The intersection of this new edge with an existing edge creates a Voronoi vertex that divides both edges and connects them together. The portion of the new edge that extends beyond this intersection, into another region, is discarded. Then, another Voronoi edge is created that bisects p_3 and the remaining (nonclosest) point, and the process is repeated. The region around p_3 is now completely defined by the new Voronoi edges that have been formed. The last step is to remove any portions of edges that are wholly contained within the newly formed region around p_3.

This process is repeated for all points in the set of points. The Voronoi edge between each point and its closest neighbor is formed first. Then, starting at the Voronoi vertex that results from the intersection of the new edge and an existing edge, additional neighboring points are evaluated clockwise based on which regions are adjacent to the new point. Voronoi edges and vertices are continually added until the entire region surrounding the new point is enclosed.

The divide-and-conquer approach behaves similarly, but instead of continually adding points, it recursively divides a set of points into multiple sets. Any given set of points could be split by a vertical line, a horizontal line, or any arbitrary line. Each subset of points will then have its boundaries established, until each subset consists of just one point. Finally, the subsets of points will be recombined into a cohesive diagram.

In general, the sweep line approach is most often used due to its straightforward implementation and optimal computation complexity. This approach is based on Fortune's algorithm, which is described in detail elsewhere [5]. At a high level, Fortune's algorithm maintains two lines: a beach line and sweep line. The sweep line can be imagined as a vertical line sweeping from left to right across the points that will compose the VD. As the sweep line encounters a point, it is incorporated into the VD via the beach line. The beach line is a curved, piecewise line that represents the current state of the VD and grows behind the sweep line as new points are added.

Voronoi skeleton generation

While VDs are typically generated using points, it is possible to adapt them for use with polygons. Such an adaptation and its optimizations are detailed in work by White et al. [6]. For the comprehensiveness of this book, the basic process as described in that work will be summarized here. When a VD is created for a polygon, the input for the generation of the VD is the set of segments that form the boundary of the polygon. For open-loop toolpath generation, the bounding edges of the remaining layer polygon are used as input. The Voronoi edges that define the regions around each segment become Voronoi skeletons.

In general, a skeleton segment will be created for every vertex of the polygon, but these segments must be pruned. Only segments fully contained within the polygon should be kept, while any segments that intersect the boundary polygon should be discarded. This simple logic allows the creation of a Voronoi skeleton that represents the remaining geometry. An example of the result produced by this VD generation method can be seen in Fig. 8.4. The dashed lines in Fig. 8.4 represent segments that were generated by the VD but are undesirable.

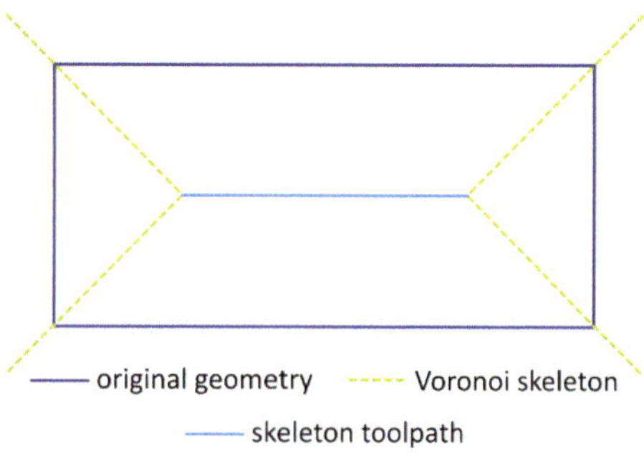

FIG. 8.4 Skeletons generated from VD overlay. All skeletons touching the boundary are pruned from the complete skeleton leaving only the skeleton toolpath (red).

Once the necessary skeleton segments have been generated, the final step is to consolidate the skeleton segments into a set of paths. Part of this path generation is to impose a specific ordering on the segments, as their generation does not guarantee an order. As such, the segments are analogous to an undirected graph. An undirected graph is a set of vertices with connecting lines, called edges, that are bidirectional. To determine a path through an undirected graph, or in our case a VD skeleton, it must be decomposed into reasonable subsets. Generally, the decomposition should result in a minimal number of paths, accounting for travel moves that must be inserted between each path. Travels were mentioned in the previous chapter and will be discussed in detail in *Chapter 11: Travels, optimizations, and ordering*, but at a high level, travels are motions where the system is moving to reposition itself, without any other action. No construction occurs during these motions, so they are generally seen as wasted time. Most approaches attempt to minimize travels or even eliminate them whenever possible.

Returning to the graph analogy, if the segments can be considered an undirected graph, then the minimal number of paths is equivalent to the minimum number of Eulerian trails that cover the graph. An Eulerian trail is a path through the graph's vertices such that every edge is traversed exactly once. Such a trail may be open or closed. Under certain conditions, it is possible to have only a single trail in a graph, specifically if (and only if) the graph contains either zero or two vertices of odd degree. The degree of a vertex is the number of edges connected to it; a vertex with an odd degree has an odd number of edges. Practically, this is unlikely for a VD of a remaining layer polygon, so there are usually multiple trails through VD skeleton segments.

To compute these trails, a modified version Hierholzer's algorithm is used [6,7]. The general logic of this algorithm is as follows:

1. Begin at a vertex of odd degree and randomly walk unvisited edges. Each time an edge is chosen, label it as visited and reduce the degree of the two vertices it joins by one. Continue in this way until encountering a vertex of degree one (an endpoint). This sequence is a candidate trail that is then set aside.
2. Repeat the previous step with the remaining vertices until there are no more vertices of odd degree; only vertices with an even degree remain.
3. Check for unvisited edges. If unvisited edges remain, it is true that a vertex must exist that is located on both a visited and unvisited edge. Any candidate trail containing that vertex is selected, broken at that vertex, and a new random walk initiated.
4. Repeat step 3 until no edges remain unvisited.

It should be noted that certain candidate walks will likely need to be prioritized during Step 1 of the algorithm. How these walks are prioritized is application-dependent, but a good rule of thumb is to avoid breaking longer walks if two or more walks contain a vertex on an unvisited edge. The reasoning for this is as stated previously: the minimization of the number of travels that will ultimately be inserted. Fewer, longer paths is generally preferrable to more, shorter paths. With this algorithm complete, a basic set of skeleton paths is generated. However, these paths may not have the necessary quality for construction. To increase the robustness of this approach, two preprocessing steps can be applied to both the generating geometry (the remaining layer polygon) and the skeleton segments: smoothing and chamfering.

Skeleton optimization

To improve the quality of the skeletons produced by the previously described process, two additional steps can be introduced. The first one is smoothing. Smoothing should be applied to the geometry used to generate the skeletons, the remaining layer polygon, and may also be applied to the skeletons themselves. In all cases, the smoothing algorithm applied is the same as the one outlined in *Chapter 5: Cross-sectioning*. The only minor difference would be the tolerance of the algorithm, which is application-dependent. Smoothing both the polygon and the skeleton segments addresses two different issues.

The first issue arises as the result of the erosion-dilation process used to construct closed-loop paths. By the time the enclosing geometry has made it to the skeleton generation step of the slicing process, it has undergone several erosions and dilations. As part of those procedures, it is possible that small irregularities exist in the remaining layer polygon. An example of this is a small deviation shown in Fig. 8.5. Any small irregularity in the polygon will cause additional unwanted skeleton segments to be generated. Though these skeletons segments would be pruned, by smoothing the polygon, they can be eliminated before the skeleton generation step.

The second issue arises with polygons that have many vertices. Such a polygon could be the result of a high-resolution STL or a geometry with a significant amount of curvature. In these cases, the VD generation will produce many small skeleton segments. Many small motions are unwanted and reduce the build quality. By smoothing the skeleton as part of the generation process, unnecessary motions can be removed.

In addition to smoothing, chamfering can also be applied to improve the quality of skeletons. Chamfering is the process of inserting a transitional edge between the two lines of a corner, to create a bevel. In the context of toolpath generation, chamfering is used on vertices that comprise an acute angle. When a VD skeleton bisects the vertex of an acute angle, it fails the skeleton pruning logic described in the previous section and gets removed. However, it needs to be included in the skeleton to capture the

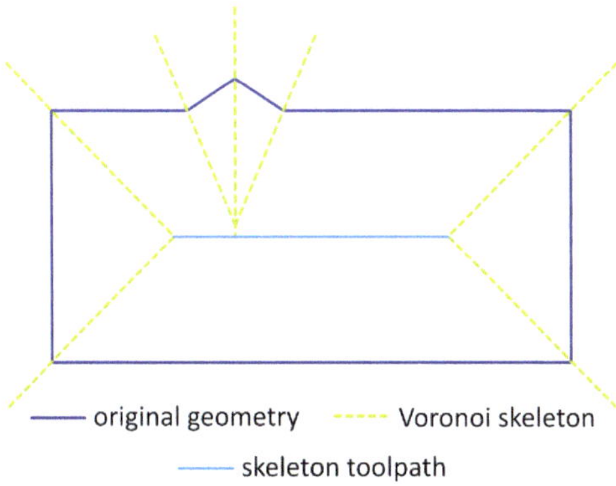

FIG. 8.5 Minor irregularity in perimeter of an object results in additional skeletons paths. In this case, the additional path could be pruned as normal, but that is not always the case. Smoothing could have removed this and returned the evaluation to Fig. 8.4.

feature of the acute angle and create a better topological approximation. To address this, vertices with acute angles can be chamfered. The chamfering splits the vertex into two temporary vertices to create a bevel. By doing so, the VD generates skeleton segments in a Y shape. The two smaller segments will be pruned, and the process will leave the one desirable skeleton segment. This process is illustrated in Fig. 8.6.

This manipulation is straightforward to carry out if the vertex can be split by a distance less than the tolerance of the machine in question. For example, splitting the vertex and separating them by a few microns is perfectly fine for most systems. Once the necessary pathing has been generated, the temporary vertices used in the bevel are discarded and

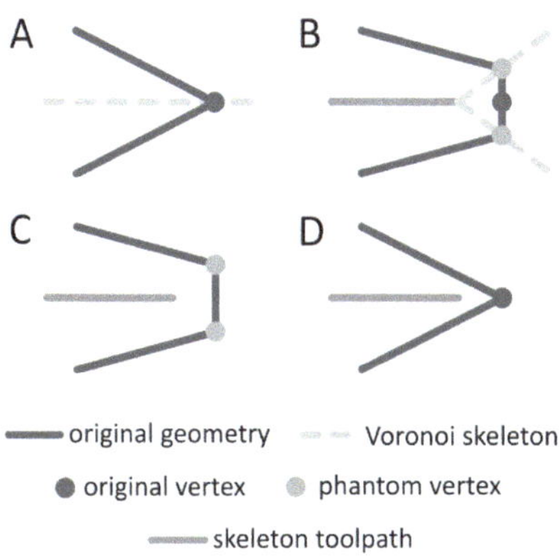

FIG. 8.6 An illustration of chamfering; (A) the original skeleton is pruned, indicated by the dashed line; (B) generation of chamfer phantom vertices and pruning of the resulting skeletons; (C) the result of the skeletal pruning; (D) the final, pruned skeleton applied to the original geometry.

the process continues with skeleton generation as normal. With this final step, the pathing can finally be connected via travels in the same way that the other path types are, which is outlined in *Chapter 11: Travels, optimizations, and ordering*.

One final note about skeletons. Because skeletons are open loops, it is much easier to adapt their bead width to appropriately fill voids. Understanding the relationship between the bead width and user settings even allows adaptive behavior such that the skeleton can expand or shrink opportunistically to fill void space. Such a process is advanced and will not be detailed here, however, interested readers can reference the cited work by White et al. [6].

Conclusion

In this chapter, an approach for the generation of open-loop paths was presented. These open-loop paths, sometimes called skeletons, are useful for filling voids that other path types cannot address. There are multiple approaches for computing these paths, but this chapter presented the use of Voronoi diagrams. Combined with pruning, consolidation, smoothing and chamfering, this approach creates robust skeletons that accurately represent the necessary portions of the geometry. The open-loop paths produced from these algorithms can be combined with the path types from the previous chapters to produce a complete pathing solution. However, there are additional path types that could be required depending on the geometry of the object to be printed and the process being used. These additional path types are discussed in the next chapter.

References

[1] H. Blum, A transformation for extracting new descriptors of shape, in: W. Wathen-Dunn (Ed.), Models for the Perception of Speech and Visual Form (PDF), MIT Press, Cambridge, Massachusetts, 1967, pp. 362–380.

[2] F. Aurenhammer, Voronoi diagrams—a survey of a fundamental geometric data structure, ACM Comput. Surv. 23 (3) (1991) 345–405.

[3] A. Sydorchuk, Boost, Polygon Voronoi Library, 2012.

[4] R. Klein, Concrete and Abstract Voronoi Diagrams, volume 400 of Lecture Notes Comput. Sci., Springer-Verlag, 1989.

[5] S. Fortune, A sweepline algorithm for Voronoi diagrams, in: Proceedings of the second annual symposium on Computational geometry, 1986, August, pp. 313–322.

[6] White, L., Quaife, B., & Borish, M. in press A novel approach for adaptive skeleton toolpath generation. Manuf. Lett.

[7] C. Hierholzer, C. Wiener, Ueber die Möglichkeit, einen Linienzug ohne Wiederholung und ohne Unterbrechung zu umfahren, Mathem. Ann. 6 (1) (1873) 30–32.

CHAPTER

9

Support, raft, brim, and skirt pathing

Alex Roschli[a], Michael Borish[a], Abby Barnes[a], Eric MacDonald[a,b], Peter Wang[a], and Thomas Feldhausen[a]

[a]Manufacturing Demonstration Facility, Oak Ridge National Laboratory, Knoxville, TN, United States, [b]University of Texas at El Paso, El Paso, TX, United States

Abstract

The previous chapters discussed the primary path types in additive manufacturing: perimeter, inset, skeleton, skin, and infill. In addition to those distinct primary path types, regions of primary paths can be assigned secondary path types: support, raft, brim, and skirt. Support paths are used to enable the printing of features that would otherwise have nothing underneath to print atop, such as overhangs. Raft paths are used to level the build plate and provide a flat surface on which to construct the object. Brim toolpaths are used to increase the footprint of the object to improve bed adhesion and reduce first layer delamination. The last path type, the skirt, is used to prime the extruder at the start of the print. These four types of path regions are created using a combination of closed-loop, open-loop, and space-filling toolpaths. This chapter will discuss how to find and generate the pathing for each region.

Introduction

The preceding three chapters have covered the main strategies for finding path locations: closed-loop pathing, open-loop pathing, and space-filling pathing. The same chapters also covered the primary path types: perimeter, inset, skeleton, skin, and infill. These path types are the only paths necessary to print an object, but they can sometimes be insufficient for a successful build. Additionally, secondary pathing regions can be added to help improve the printing process and ensure a successful build. These additional regions are support, raft, brim, and skirt. They each

https://doi.org/10.1016/B978-0-443-15286-3.00015-1

employ a combination of the path-finding strategies and path types covered in the previous chapters, and each is used in a particular print scenario that may not occur in every print. The following sections of this chapter will explore each of the four regions, including when to use them, how they are applied, and examples of their implementation.

Supports

What happens when the geometry of the printed object changes such that the current layer polygon is larger than the previous layer? The extrusion process cannot overcome gravity and float extruded material in the air; the current layer must either print directly above the previous layer or additional toolpaths must be created underneath it. This additional region of toolpaths is called "support." Support toolpaths are sacrificial toolpaths printed specifically to allow for printing of unsupported overhangs and bridges. At the completion of a print, the support material can be manually removed and recycled, or it can be dissolved away if the material is dissolvable.

There are two types of unsupported paths: bridges and overhangs. Bridges are unsupported areas that are connected on at least two opposing sides such that a single path can be generated to "bridge" the gap between the supported areas. Bridges don't always require support structures, depending on their length and 3D-printing process in use. Fig. 9.1 shows an object with a bridge.

Overhangs are unsupported areas that are cantilevered and, unlike bridges, are not connected by two opposing sides of the supported geometry. They can occur when a large area that was not present on any preceding layers is suddenly present on a layer (Fig. 9.2A) or they can occur when the object slowly builds outward layer after layer (Fig. 9.2B). For overhangs that gradually build outward, layer after layer, the angle at which the object builds out, called the overhang angle, determines whether the print requires support structures.

FIG. 9.1 A printed object with an unsupported bridge area.

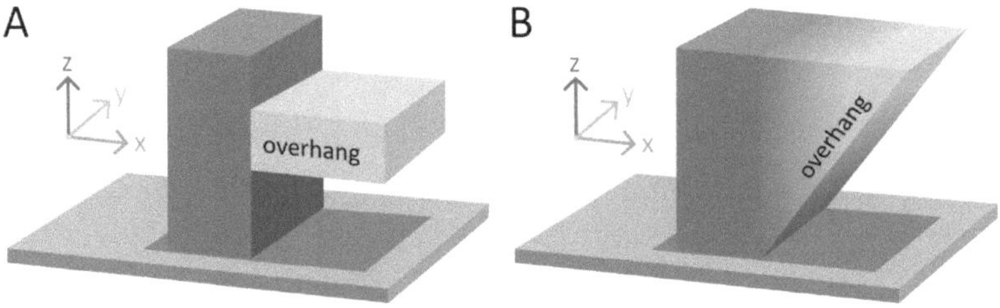

FIG. 9.2 (A) an unsupported overhang and (B) an overhang showing the transition from self-supporting to unsupported.

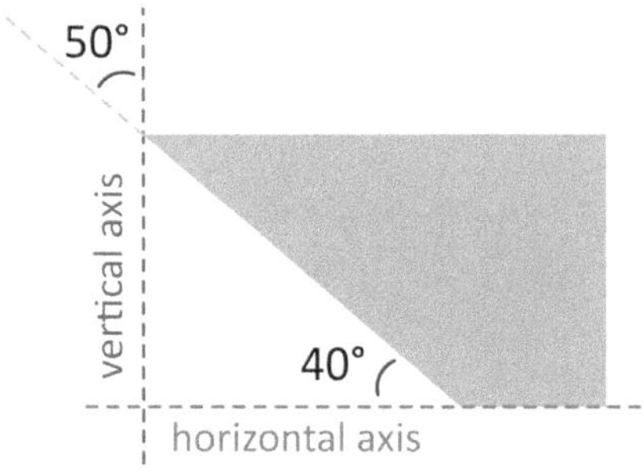

FIG. 9.3 One overhang with two different measurement references.

When talking about overhang angles, it's important to distinguish the reference axis of the measurement. There are two standards: vertical and horizontal. Measurements from the vertical axis start with 0 degrees, no overhang at all, and go up to 90 degrees, completely overhanging the previous layer. Measurements from the horizontal axis start with 0 degrees, completely overhanging the previous layer, and go up to 90 degrees, no overhang at all. Fig. 9.3 shows an example overhang with both the horizontal and vertical measurements given.

The first step to generating support structures is finding the areas of the object being printed, called the "primary object," that need support. These areas are programmatically found based on the user-defined "support angle," which is the threshold that determines the range of overhang angles that require support structures. For example, a horizontal support angle of 40 degrees indicates that any overhang less than 40 degrees from the horizontal axis requires support structures. Like the algorithm used for finding skins in space-filling toolpath generation, the process of finding areas that need support structures starts with a polygon comparison between the current layer polygon and the previous layer polygon. In areas that meet the overhang threshold, the support angle is used to oversize the previous layer polygon just for comparison purposes (the polygon is not permanently oversized). The amount of oversize is found using the

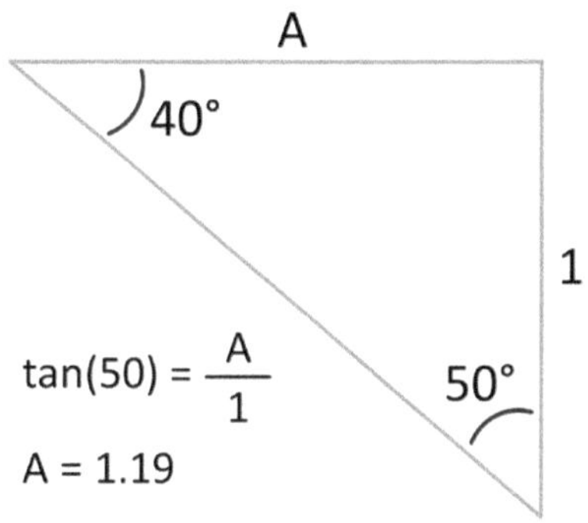

FIG. 9.4 Calculating the oversize area that doesn't need support, A, based on the layer height and angle.

tangent function of the vertical overhang angle. For a layer height of 1, with a horizontal support angle of 40 degrees, the vertical support angle is 50 degrees and the tangent of 50 gives a result of 1.19 (Fig. 9.4). This means the layer will be oversized by offsetting the layer polygon outward by 0.19 (i.e., enlarging by 19%).

If the area from the current layer polygon fits entirely within the bounds of the oversized previous layer polygon, support structures are not needed. If there is area from the current layer that does not overlap the oversized previous layer, that area will require supports. The area that requires support is captured as a polygon and added to the previous layer. Then, the previous layer, where a support polygon was added, runs through the same comparison algorithm with the layer below it to determine where support structures are needed.

After comparing all of the layers for a part, the resultant support polygons are ready to be fit with pathing. Because support material is always sacrificial, whether it is manually removed and discarded or dissolved away, it's always ideal to minimize the amount of printed support to shorten print time and lessen material use. Support structures are therefore often built with only space-filling toolpaths, typically sparse infill, except for where they meet the primary object, at which point they need to be solid. Sparse space-filling paths are also used because support structures are meant to be printed quickly and they don't require the high accuracy or tight tolerances that the primary object requires. For small support structures, a closed-loop contour can be added to bound the support structure and give it rigidity. User settings can be used to define how close the support structure gets to the primary object in the X and Y directions, called the horizontal offset, as well as how close the support gets to the underside of the object, called the vertical offset.

The area where the support material meets the primary object material is called the interface layer. This is the most important layer of the support pathing. The interface layer supports the load of the first supported layer of the primary object and ensures that the primary object paths are correctly printed. The interface layer must also separate easily from the primary object after printing, if dissolvable support material is not used. The interface layer is typically printed using a line fill pattern with a full, or nearly full, density. A vertical offset can be used to create a small gap between the interface layer and primary object to help lessen the bond between the layers. The goal is to provide a structure for the primary object to rest on, but not create a layer-to-layer bond.

Sometimes, an interface layer doesn't only occur at the top of the support structure, where the primary object is printed over the support. When support pathing originates on top of the primary object, an interface layer can also occur at the bottom of the support structure (Fig. 9.5). In this instance, the support structure begins with an interface layer and transitions to sparser fill when moving away from the interface layer.

As previously mentioned, it's always ideal to minimize the amount of printed support to save time and material. One way to do this is tapering support structures that shrink in cross-sectional area as they move away from the primary object. A tapering support structure can be seen in Fig. 9.6A. Another, more complex approach is "tree support," which uses organic shapes to create support structures that minimize material [1]. Fig. 9.6B shows an object with tree supports.

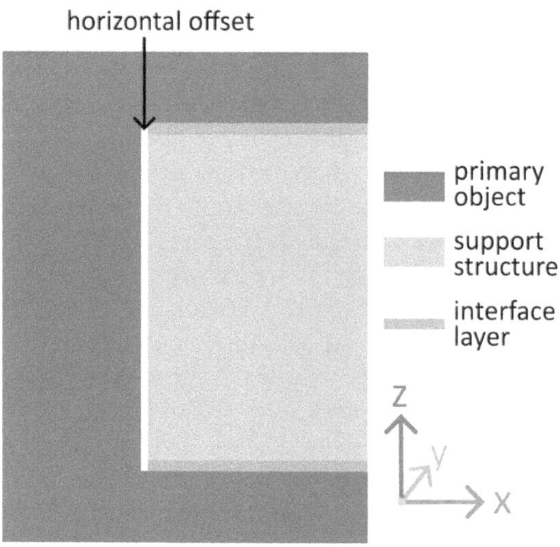

FIG. 9.5 An object with support structures that grow on top of the object itself.

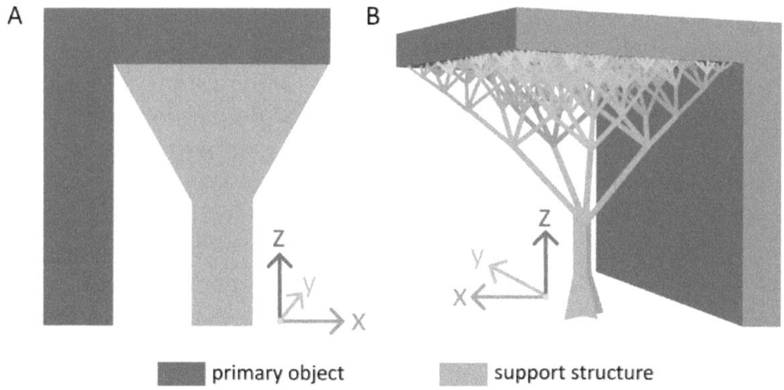

FIG. 9.6 (A) Tapering supports and (B) tree supports.

Rafts

For most AM processes, printing begins by depositing a first layer onto a flat sheet. The success of that first layer can often be directly correlated to the flatness of the sheet. A sheet with holes, high and low spots, or that is tilted, can prevent a first layer from properly sticking and can disrupt subsequent layers. The best approach is to use a clean and flat sheet that is properly leveled at all corners so that the first layer prints successfully. However, for systems using a small first layer height, it may not be achievable to get the bed level enough for a successful first layer. For these systems, a special print toolpath, called a raft, can be applied. A raft functions like a support structure for the object's entire first layer by providing a flat, even foundation for the whole layer.

A raft is printed beneath the object by offsetting the object away from the build surface. The raft is usually printed with different settings than the primary object, such as a thicker layer height. A larger layer height is less sensitive to inconsistencies in the levelness of the bed, which is ideal for printing the first layer when the bed may not be level to the motion of the gantry. For example, a 1 mm layer height can only overcome levelness inconsistencies up to ~0.5 mm but a layer height of 3 mm allows for inconsistencies up to ~1.5 mm. It's for this reason that larger printers, which output a larger bead of material at a thicker layer height, are less likely to require raft toolpaths. Fig. 9.7 shows a raft printed on top of a non-flat build surface to create a flat surface for the object to be printed.

The raft region is found by using the first layer polygon of the object and offsetting it outward by a user-defined distance, called a "raft offset," that determines how far the raft should extend from the object. This raft polygon can then be fitted with toolpaths, which is typically done entirely with space-filling toolpaths and no closed- or open-loop contours. Like support

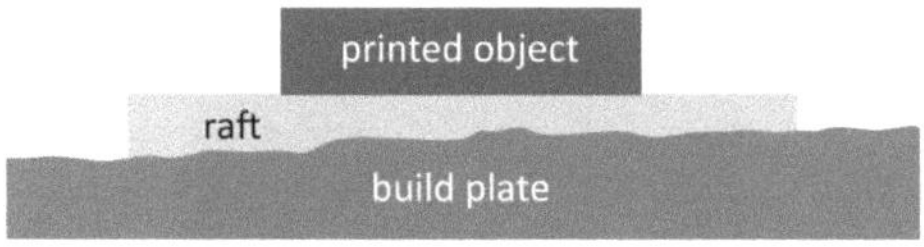

FIG. 9.7 A raft structure printed before the primary object to account for a non-flat build plate.

pathing, the pathing for each raft layer varies by proximity to the primary object. The last layer of the raft (i.e., the top layer) is the interface layer, which needs to be printed with a high enough fill density to support the object's first layer. All other layers of the raft can be printed with a less dense fill to save material and print time. The user can define the fill density and fill pattern, but it's most common to use a rectilinear fill because it is easy to print on top of and prints quickly.

Large objects can often be very difficult to separate from the raft because the increased surface area creates a strong bond between the raft and the primary object. Like support material, the raft is often printed with a material different from the primary object because it allows for easier separation or dissolving of the raft.

Brims

3D-printing processes create a thermal gradient from layer to layer during the printing process. As a layer is deposited while hot, it slowly cools to the ambient temperature over time, while additional hot layers are added on top. Most metal and polymer materials used in 3D-printing expand when hot and contract when they cool [2,3]. This can cause the object being printed, particularly the first layer, to warp away from the build surface, which causes the print to fail. This effect is especially prevalent on sections with thin walls and objects with a small first layer footprint.

A brim region is one way to combat the warping and delaminating of a first layer. A brim is used to increase the surface area of the first layer, which helps to increase bed adhesion by increasing the bonding area between printed object and substrate. The brim is an expansion of the first layer, or layers, using closed-loop contour paths to offset the object outward. Fig. 9.8 shows a cube shape with a brim surrounding it.

To find the brim region, the first layer polygon is offset by a full bead width, as discussed in *Chapter 6: Closed-loop toolpath generation*. However, in the context of brims, the offset is outward rather than inward. This new polygon is then offset inward by one half bead width to find the centerline, where the toolpath will reside. This process is repeated until the brim region is sufficiently wide and all the brim toolpaths are found. The

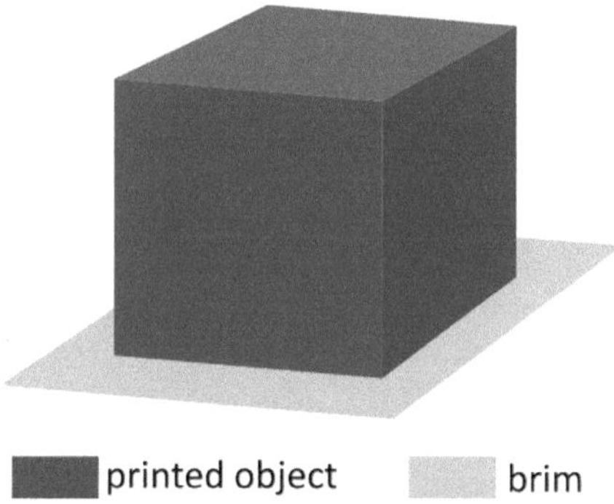

FIG. 9.8 Brim pathing surrounding a cube.

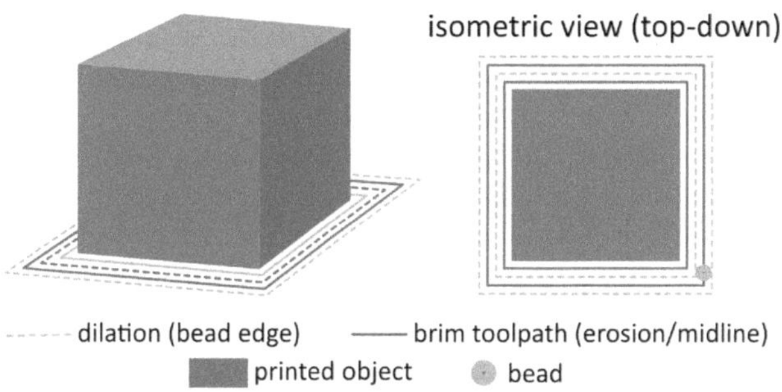

FIG. 9.9 Brim toolpath locations using erosion and dilation.

amount of brim can be defined by a distance value that defines the brim width, or by an integer number of beads, both of which can be user settings. When the user defines the brim width, the number of closed-loop toolpaths is found by dividing the brim width by the bead width. Fig. 9.9 shows the brim toolpath locations for Fig. 9.8.

For complex shapes, the outward offset process could cause the offset path to intersect itself or even the primary object. In both cases, the primary object's layer polygon takes highest importance, and its shape will not be altered. The brim polygon will get as close to the object as possible and create as many toolpaths as the space will allow without intersecting the object. Additionally, just as discussed in *Chapter 6: Closed-loop toolpath generation*, areas in which the offset path intersects itself are considered negative area and are removed from offset polygon, and therefore do not get carried over to the brim toolpath. Fig. 9.10 shows a polygon with different size gaps and illustrates negative area that the brim pathing ignores.

When multiple parts are being printed at one time, the brim pathing is generated for each object individually. Like the example of one object with

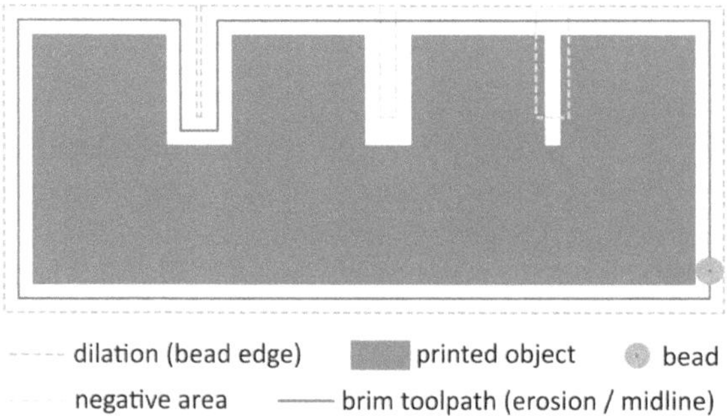

FIG. 9.10 Brim generation process for an object's first layer including areas that are too small for a brim.

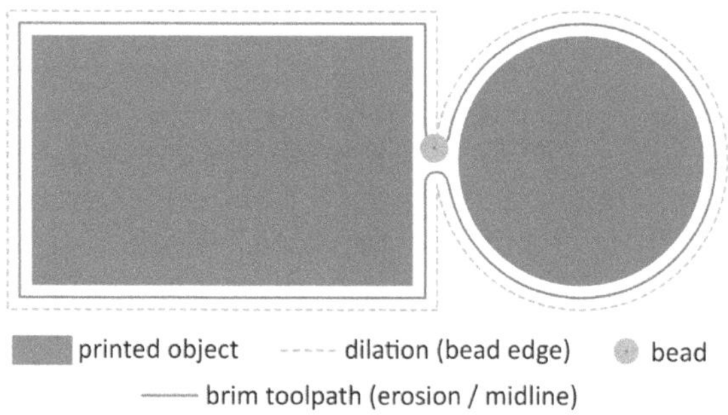

FIG. 9.11 Merging of brim paths for two distinct objects.

a complex shape, the brim pathing for multiple objects can grow large enough that the pathing intersects. If this happens, the brim regions are joined together. Fig. 9.11 shows an overhead view of a layer where two objects have brim paths that merge. Some slicing programs will allow brim pathing to only be generated for specific objects, as defined by the user.

The brim can be applied to only the first layer or can be applied to multiple layers to ensure a sufficient bond between brim and primary object. Larger objects take more time to print and cause more residual stress to build up because the thermal gradient between layers becomes more significant. Because of this, it can be useful to print the brim on multiple layers so that the brim doesn't peel away from the part during printing. The downside to using multiple layers and creating a better bond between primary object and brim is that the brim can become so well bonded that it is difficult to separate after printing.

While the brim is typically used for improving the first layer bond, it can also be used as a means of priming the extruder. This is particularly useful for prints that don't involve an initial purge or skirt pathing, which will be discussed in the next section.

Skirts

The standard order of operations for beginning a print involves preparing the machine and workspace, loading the print material, heating the system (for thermal processes), and finally priming or purging the extruder. The purging process includes removing old material left in the extruder from the previous operation as well as starting the flow of the new material to achieve a steady state. Some systems have a designated location for purging material into a bucket or waste bin, but this requires some method of clearing the purged material from the nozzle tip before starting the print. This may involve a wiper or a human with a spatula to scrape it off. The human method, as well as the scraping method, can induce a pause between priming the extruder and beginning the print. For screw-based pellet extruders, low viscosity resin extruders, and other systems that have no means of retraction or nozzle shutoff, this pause leaves time for material to ooze out of the extruder before starting the print. This material that oozes out, called a "tail," stays attached to the nozzle and becomes part of the start of the print.

An automated method of priming the extruder that helps minimize tails is called a "skirt" toolpath. The skirt is a sacrificial print path that is implemented for the purpose of priming the extruder. When skirt pathing is added, it is the first pathing to be printed on the layer. Unlike brims or support structures, a skirt is not meant to support the primary object in any way and therefore does not touch or bond to the object, so it doesn't have to be broken away.

A skirt toolpath makes at least one closed loop around the object. Skirt toolpath locations are typically found by means of a convex hull or bounding box. The convex hull approach uses an algorithm to create a simplistic, convex representation of the geometry by removing any concavity; the term "convex" indicates that a line drawn between any two points on the edge of the representation will not intersect the edge at any other points. Fig. 9.12A and B show a 2D shape and its corresponding convex

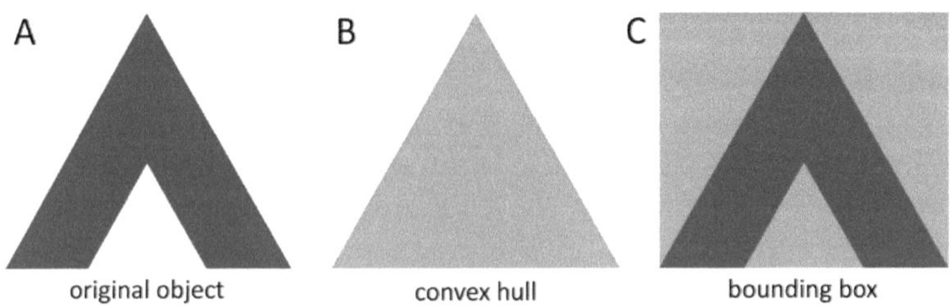

FIG. 9.12 (A) an original layer polygon with concavity, (B) the resultant convex hull polygon for the geometry eliminates the concavity, (C) the bounding box polygon for the geometry creates a rectangle to fully encompass the original geometry.

hull, respectively. The bounding box method works by creating a bounding box around the entire object using the minimum and maximum X and Y values for the object. Fig. 9.12C shows a bounding box created for the object shown in Fig. 9.12A.

An added benefit of printing a skirt is that it can be used as an indicator of calibration and bed levelness before starting the print. Because the skirt traverses the entire outer boundary of the layer, and is the first toolpath printed, just noting whether the skirt adheres to the print bed during all points of its toolpath is enough to understand how the first layer will stick. Some systems allow for real-time adjustment of Z height (i.e., print height) which can be used to make minor adjustments during printing of the skirt.

After the convex hull polygon or bounding box polygon has been found, the skirt toolpath can be found by offsetting the polygon outward, like how brim toolpaths are found. Unlike brim toolpaths, however, the outward offset of the polygon of a convex hull or bounding box skirt cannot cause intersections, which means complex geometries do not need any additional processing steps. The user simply defines how far to offset the skirt toolpath from the object, then this offset polygon, called the skirt polygon, is inset by half a bead width to find the toolpath location.

If the skirt toolpath is to be based on the bounding box, a polygon is drawn using the maximum and minimum X and Y values of the entire object, comparing every layer. Then, that polygon is offset outward by the user-defined distance to find the skirt polygon. By using the global maximum and minimum values, rather than a single layer's maximum and minimum values, the skirt can be built multiple layers tall without concern of intersecting the object. Figure 9.13 shows the skirt toolpaths for the object from Fig. 9.12 using both the convex hull method and the bounding box method.

FIG. 9.13 Two methods of generating the skirt around a single object: convex hull and bounding box.

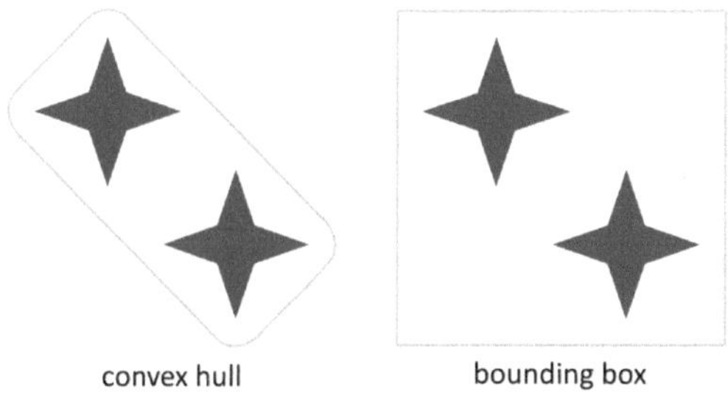

FIG. 9.14 Two methods of generating the skirt around multiple objects: convex hull and bounding box.

When multiple objects are being printed at the same time, a single skirt is generated around all objects. The skirt polygon can be found using the same two methods, either convex hull or bounding box. The convex hull approach works like the previous scenario with a single object, but all the points of all the objects are used by the convex hull algorithm to create one convex shape that encompasses all of the objects. The bounding box approach works by finding the global maximum and minimum for all objects, then creates a box with these points. After the convex hull or bounding box has been found, the resultant polygon can be offset outward to generate the skirt toolpathing. Both example toolpaths can be seen in Fig. 9.14.

Additional user settings for defining skirt pathing include minimum length and number of layers. For smaller objects, one single closed-loop path around the object may not be enough to fully prime the extruder and reach a steady state of material flow. The minimum length setting allows the user to define a minimum print distance for the skirt region. If one closed loop isn't enough for the minimum length, additional closed-loop paths can be added.

The number of layers is a user setting that allows the skirt to be continually printed after the first layer. This can be helpful for checking the flow from layer to layer, by comparing the output from the first layer of the skirt to subsequent layers. The number of layers can also be set to the full height of the object, so that the skirt is printed on every layer. This is useful as a draft shield, to protect the printed object from drafts of air or nearby fans that could undesirably increase the cooling rate during printing. The taller skirt also helps to trap warm air around the object to help slow the cooling rate. Cooling too quickly can decrease the layer-to-layer bond strength, which can cause delamination or warping [4]. Fig. 9.15 shows a cylinder with a full height draft shield.

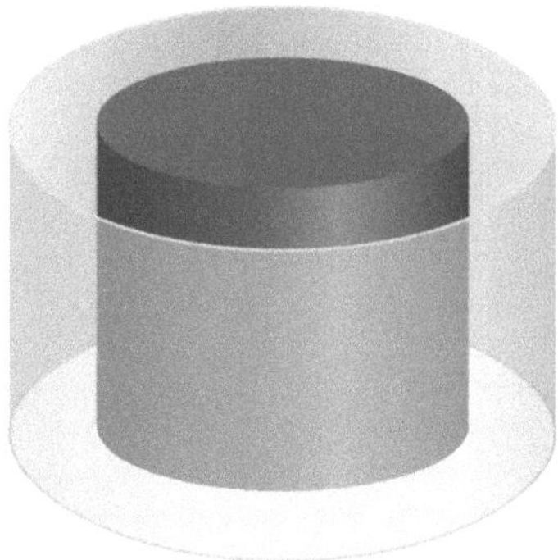

FIG. 9.15 A cylinder object (purple) with a draft shield printed all the way around (red).

Conclusion

Many different toolpath types exist for the creation of a 3D-printed object. The primary path types (perimeter, inset, skeleton, skin, and infill) are used to build the object and ensure the volume is filled according to the design. However, these path types can also be used in four secondary toolpath regions: support, raft, brim, and skirt. These regions are used to enhance the printing process and ensure a successful print. A support region is used to enable printing of bridges and unsupported overhangs. A raft region helps overcome build plate inconsistencies by providing a flat foundation for object construction. The brim region helps keep the object's first layer flat by increasing surface area, to eliminate warping. Lastly, the skirt region is helpful for priming the extruder and can be used to block ambient air from cooling the print. While each secondary region may not be used on every print, they all have a specific purpose and can be applied as needed to ensure successful printing.

References

[1] N. Zhang, et al., Local barycenter based efficient tree-support generation for 3D printing, Comput. Aided Design 115 (2019) 277–292.
[2] C. Greer, et al., Introduction to the design rules for metal big area additive manufacturing, Addit. Manuf. 27 (2019) 159–166.
[3] C. Duty, et al., What makes a material printable? A viscoelastic model for extrusion-based 3D printing of polymers, J. Manufac. Proc. 35 (2018) 526–537.
[4] E. Polyzos, et al., Delamination analysis of 3D-printed nylon reinforced with continuous carbon fibers, Addit. Manuf. 46 (2021) 102144.

CHAPTER

10

Path modifiers

Alex Roschli[a], *Michael Borish*[a], *Abby Barnes*[a], *Thomas Feldhausen*[a], *Eric MacDonald*[a,b], *and Peter Wang*[a]

[a]**Manufacturing Demonstration Facility, Oak Ridge National Laboratory, Knoxville, TN, United States,** [b]**University of Texas at El Paso, El Paso, TX, United States**

Abstract

Toolpaths for additive manufacturing are typically broken into three types: closed-loop, open-loop, and space-filling. Each of these toolpaths serve a specific purpose and are often used in combination to construct an object. The standard implementation of each toolpath commands a constant motion speed and extrusion rate throughout the length of the path. The 3D printer will then make any necessary adjustments during interpolation of the motion, such as accelerating and decelerating when traversing a corner. However, sometimes more finite control and additional pathing can be needed to ensure the successful printing of a path. This chapter will outline path modifiers and how they can be used to enhance the printability of a toolpath by modifying parameters or adding segments.

Introduction

The previous chapters have covered the process of finding toolpaths for the primary and secondary regions of a print. This includes finding the location for a path, fitting the path to the area it will cover, and finally planning the path's points and trajectory. For any given path, the result is a list of points that the 3D printer will sequentially traverse to create the path. This list of toolpath points can now be written out to the g-code file, which is discussed in *Chapter 18: The G-code file*. However, before writing the toolpath, there is an opportunity to modify or enhance the path to add functionality, improve the quality, and/or ensure a successful print. Such changes to a path are termed "path modifiers." Path modifiers are modifications made to toolpaths that have already been planned. In fact, path

https://doi.org/10.1016/B978-0-443-15286-3.00018-7

modifiers are added after *all* toolpaths on a layer have been planned. The following sections of this chapter will cover various path modifiers including how they work, when they can be used, and how they are planned.

Tip wipe

At the conclusion of a path, regardless of path type, the extruder is turned off, and the system follows a travel path to move to the start of the next path and begin extruding again. However, at the end of the current path, when the extruder turns off, the material in the extruder's nozzle is still connected to the material extruded onto the layer. This material needs to be broken away from the layer cleanly, so that material is not left protruding from where the path ended. Another concern, particularly with extruders that operate under pressure, is that material may still be flowing or oozing out of the nozzle. Most extruders work by creating pressure inside the barrel and using that pressure to extrude material through a nozzle. When the extruder turns off, the pressure inside the extruder is released by forcing additional material out of the nozzle [1].

To break the bead off at the end of a path, and also to compensate for the pressure release that causes material to ooze from the nozzle, a path modifier is needed to alter the ending of the extrusion path. This path modifier, called a tip wipe, is used to wipe the nozzle off after an extrusion so that the nozzle is clear of material and debris. The tip wipe is inserted at the end of the path and before the travel to the next path.

Tip wipes come in three different directional varieties: forward, reverse, and angled. Each direction can be applied to the three main path types (closed-loop, open-loop, and space-filling). The implementation for closed-loop paths can vary from that of open-loop paths, but open-loop and space-filling are always implemented in the same way. For all wipe directions, the user defines the motion speed and the total length of the tip wipe, termed the "wipe distance" or "wipe length."

Of the three directions, forward tip wipes are the most complex and also deviate the most from the path planning steps discussed in the previous chapters. The concept behind the forward wipe is to continue traversing forward along the current printing vector with the intention of wiping across the top of something that has already been printed.

For a closed-loop path, a forward tip wipe continues to follow the original print path (Fig. 10.1). This is implemented by stepping in order through the points of the print path. If the wipe distance is not precisely met by traversing to an existing point of the original print path, a new point can be added so that the total wipe length matches the desired wipe length (Fig. 10.2).

To break a segment and insert a new point at length X along the segment, barycentric coordinates can be applied [2]. First, a modifying value,

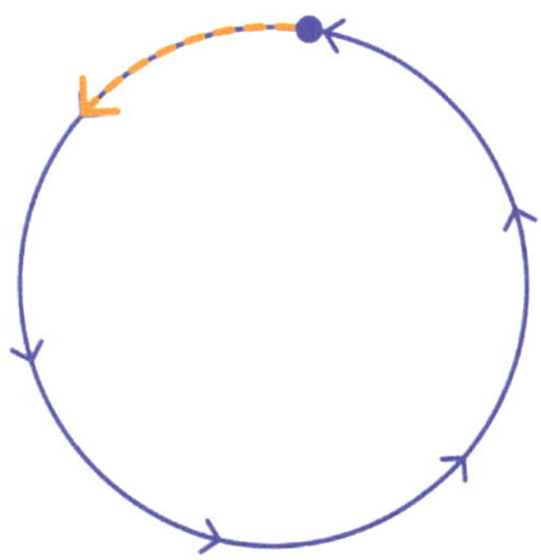

FIG. 10.1 Forward tip wipe modifier added to a circular path.

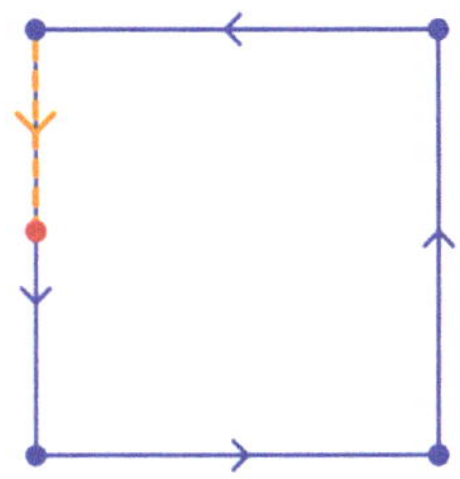

FIG. 10.2 Inserting a point to make the tip wipe length precise.

M, must be found using Eq. (10.1), where segment distance represents the full length of the segment to be broken and wipe distance represents the distance along the segment where a point should be inserted. Next, Eq. (10.2) is used to calculate the X location, P_X, of the new point using the X location of the first point of the segment, P_{1X}, and the X location of the second point of the segment, P_{2X}. Finally, Eq. (10.3) is used to calculate the Y location, P_Y, of the new point using the Y location of the first point of the segment, P_{1Y}, and the Y location of the second point of the segment, P_{2Y}.

$$M = 1 - \left(\frac{\text{Wipe Distance}}{\text{Segment Length}}\right) \tag{10.1}$$

$$P_X = (1 - M) * P_{1X} + M * P_{2X} \tag{10.2}$$

$$P_Y = (1 - M) * P_{1Y} + M * P_{2Y} \tag{10.3}$$

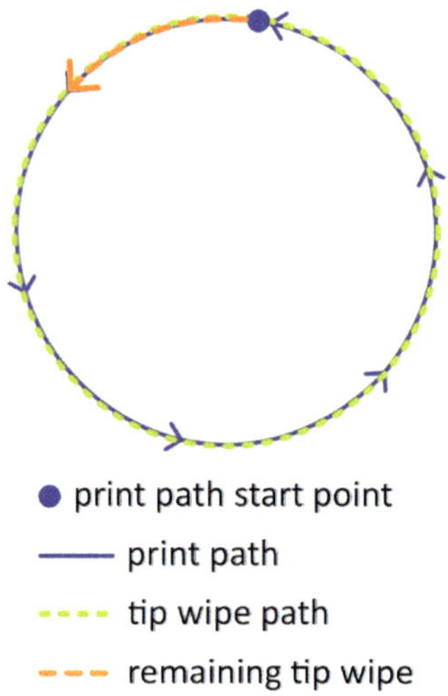

FIG. 10.3 Example of a tip with length longer than the total path's length.

If the wipe distance is longer than the total length of the print path, the tip wipe will repeat its steps through the points of the path until the total distance is achieved. This means completing one full revolution through all of the points in the print path and then continuing through points that have already been passed once. Fig. 10.3 shows an example of this scenario when printing a circle.

The approach of stepping through the points of the current path does not work for open-loop paths, because the start and stop point of the path are not the same and can be far apart. So, a different implementation must be used for open-loop forward tip wipes. Depending on the available geometry for the layer, the forward wipe can be executed as either a contour wipe or an open-loop forward wipe. A contour wipe is only possible if there is at least one perimeter or inset path that acts as a bounding contour for the current open-loop print path. The wipe is implemented by first finding the path of the contour, then finding the segment of the contour that is closest to the end point of the open-loop path. Next, a point is added along that contour segment, at the location closest to the end point of the open-loop path, and a wipe path segment is added to connect the end of the open-loop path to the newly added point. Then the remaining tip wipe segments are created by following the points of the contour path until the total wipe distance has been achieved, just like in the closed-loop forward tip wipe. Fig. 10.4 shows the result of this pathing.

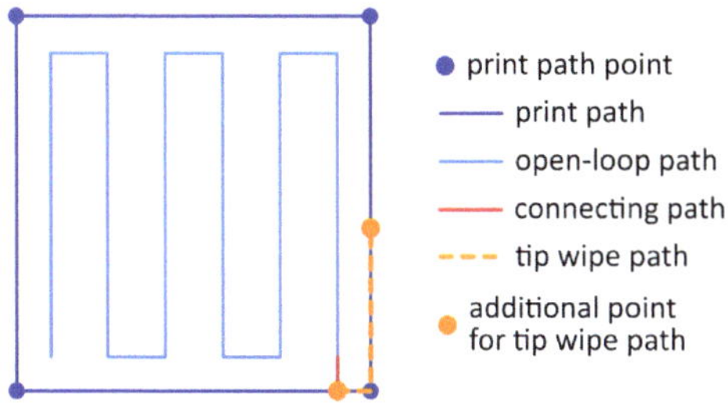

FIG. 10.4 Infill tip wipe pathing with a connecting to, then a continued path wipe over the existing contour pathing.

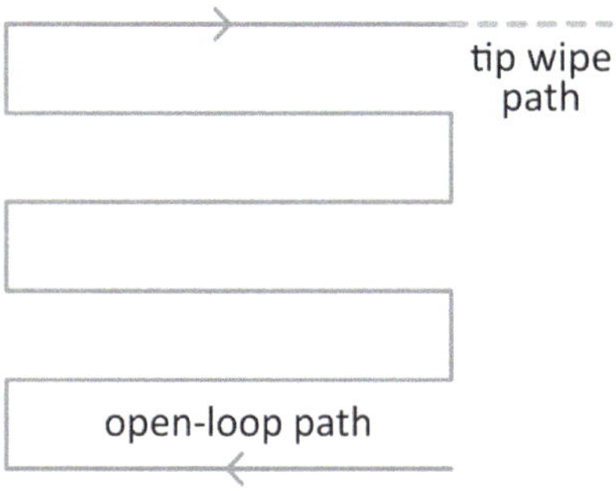

FIG. 10.5 An open-loop path tip wipe.

When a bounding contour isn't available, an open-loop tip wipe is implemented by continuing the path along its existing vector until the wipe distance is met. This wipe is the least effective implementation of the forward wipe because there is nothing to wipe the nozzle across and help break the bead. Instead, the wipe is done at a high speed to quickly stretch and break the bead free. This open-loop tip wipe can be seen in Fig. 10.5.

A reverse tip wipe is more straightforward to implement because the approach is more broadly applicable to all path types. At the end of the path, the print head reverses direction and re-traverses its path until the wipe distance has been achieved. Similar to forward wipes, it's likely that a segment will need to be broken, using Eqs. (10.1–10.3), to achieve the exact wipe distance. Fig. 10.6 shows a reverse tip wipe path.

The only difference in implementation of reverse tip wipes for open-loop vs closed-loop paths occurs during the instance of a tip wipe that is longer than the printing path length. For closed-loop paths, like with forward tip wipes, the pathing continues stepping through the points of the closed-loop path and may make multiple revolutions to meet the total distance. For open-loop paths, once the entire list of points has been stepped through,

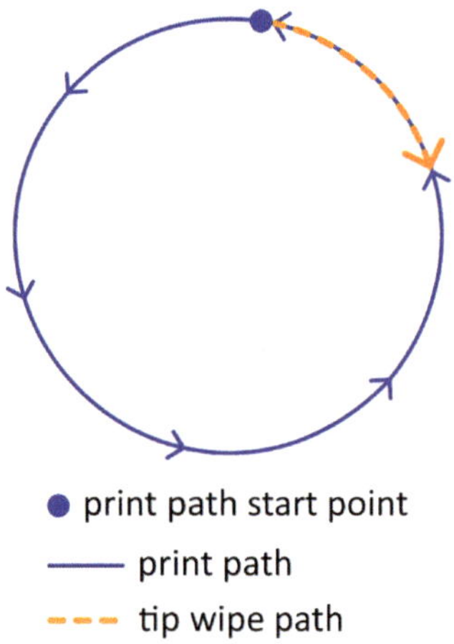

FIG. 10.6 Reverse tip wipe added to a circular toolpath.

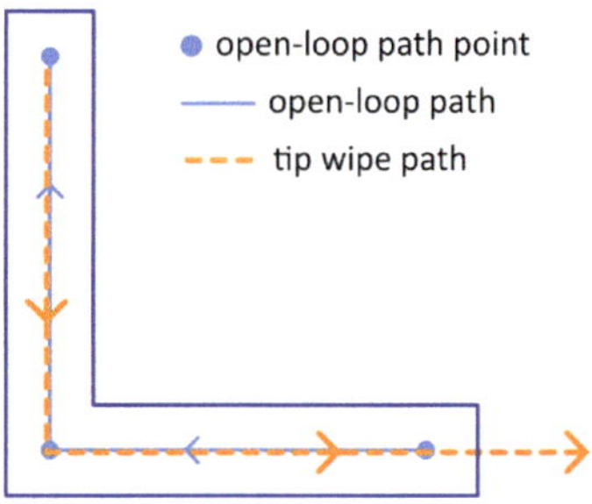

FIG. 10.7 Reverse tip wipe for an open loop that extends outside the bounds of the object.

the reverse tip wipe will continue along the last vector, creating a straight wipe path that may not travel over any printed area (Fig. 10.7).

The angled tip wipe is the simplest implementation of the three and applies identically to all path types. It requires a unique user input: a wipe angle from 0 to 359 degrees. A single path segment whose length is equal to the wipe distance is generated at that angle, relative to the current trajectory of the path, with 0 degree being in-line with the current printing direction. That's it. The entire wipe is just a single segment at a set angle from the printing direction. This wipe segment may or may not cross the top of other printed paths. The last step of this approach is to check the end point of the wipe to make sure the machine limits (min/max *X* and *Y*) have not been crossed by the wipe path. Fig. 10.8 shows an angled wipe, at 45 degrees, for a closed-loop path.

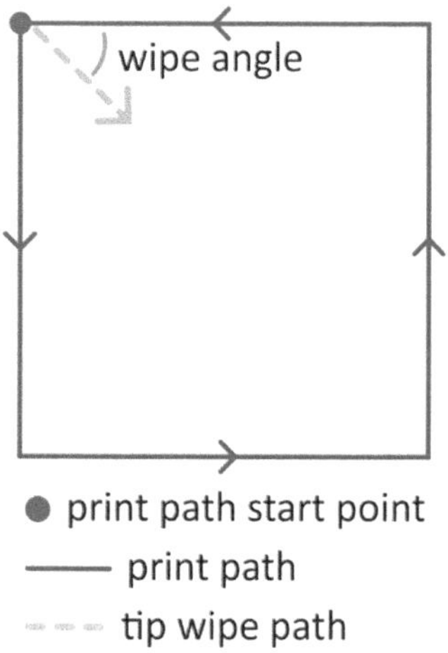

FIG. 10.8 Angled tip wipe path.

Spiral lift

The spiral lift is a path modifier that can be used as a replacement for, or an addition to, a tip wipe. Even when using a tip wipe to separate the bead from the material in the nozzle at the end of a print path, extruded material meant to be part of the object can be left with a rough surface due to the raised position produced by the vertical lift motion of a tip wipe. This is especially bad for the material directly under the nozzle during the lift. A spiral lift counteracts this by adding a spiral motion during the lifting process, which helps to flatten down a circular area surrounding the lift point so that even the lift point gets flattened. Fig. 10.9 shows a 3D view of the spiral lift motion.

A spiral lift can be implemented on any system, but it works best for extrusion heads that have a leveling device, such as a reciprocating platen or tamper, surrounding the nozzle. The spiral motion gives the leveling

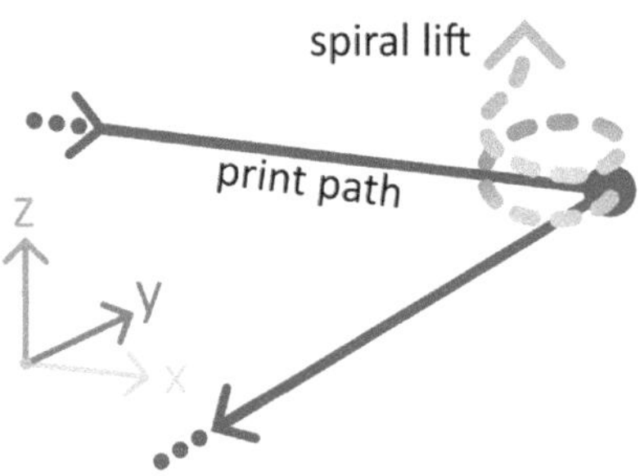

FIG. 10.9 The three-dimensional spiral lift path modifier.

device time to flatten everything surrounding the end of the bead, including the area under the nozzle at the end of the path. Without a spiral lift, the vertical motion never gives the leveling device an opportunity to level what is under the nozzle.

The spiral lift can be implemented with linear G1 motions or circular arc G2 and G3 motions. These motion commands will be discussed in *Chapter 18: The G-code file*. For both instances, the settings allow the user to define the diameter of the spiral motion, the speed, and the lift height.

Startup

As previously mentioned, pressurized extruders need time and space to release pressure as the extruder turns off. Likewise, when turning on these extruders, they need time to build pressure and replace the volume of material in the nozzle that may have been lost to oozing while turned off. At the start of the path, if the extruder immediately moves to the location and begins printing at full speed, the bead may not properly adhere or print as desired because the interior of the extruder has not reached the pressure and volume of material required to establish a steady flow state.

To combat this issue, a startup path modifier can be used at the beginning of any path type to ensure the establishment of a proper flow state. The startup works by printing the beginning of a path at a different motion speed and extrusion rate. The typical implementation scales down the speed and extrusion rate, but not by the same scale factor. The extrusion rate is usually scaled down by a smaller amount than the speed of motion so that the extruder can build pressure and start the flow of material relatively quickly. Without this startup path modifier, the bead may not adhere at the start of the path and can drag behind the nozzle until enough material has been extruded to adhere to the substrate or previous layer. Additionally, the first portion of the path may not be printed at all because the material is not immediately extruded.

A startup modifier does not add additional moves or paths to the print; it only serves to modify the existing pathing by adjusting the motion speed and extrusion rate. The startup path is implemented by modifying the first X distance of the print path, where X is a distance defined by the user in the slicing settings. To achieve the exact distance for the startup, a path segment may need to be broken into multiple parts using Eqs. (10.1–10.3). Once the startup section has been found, the resulting path segments can have their motion speed and extrusion rate set independently from the rest of the print path. Fig. 10.10 shows a square toolpath before and after the startup path has been added.

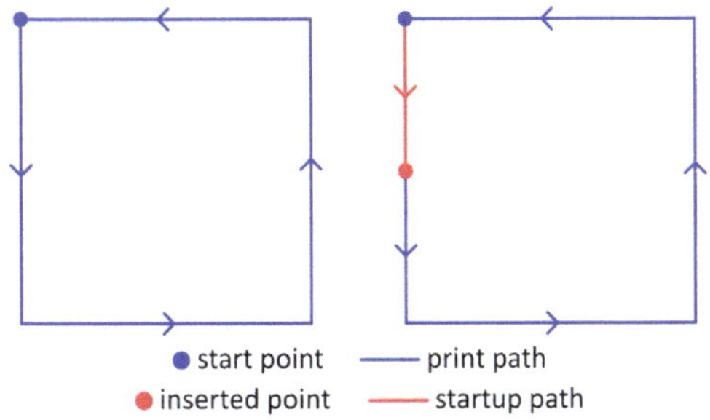

FIG. 10.10 A square toolpath without and with a startup modifier.

For some systems, a single startup "step" is not sufficient because it causes too drastic of a change in state. Adding more, smaller steps to transition from shutoff to printing speed creates a smoother transition to steady state. This is especially important for systems where the extrusion axis and printing axes do not accelerate at the same speed. The ability to add more steps is achieved with a startup "ramp up."

A ramp up allows the user to define a number of discrete steps to transition from the startup parameters (speed and extrusion rate) to the desired path parameters. This is implemented by allowing the user to define the motion speed and extrusion rate of the first step as well as the desired number of steps after the initial step. The user also defines the total length of the ramp up. The original printing path will be broken into equal-length segments, the lengths of which are found by dividing the ramp up distance by the number of steps. Then each step will be assigned a speed of motion and extrusion rate, each of which is found by subtracting the startup parameters from the original path parameters and dividing the remainder by the total number of steps (including the initial step). The segments of a startup path with a ramp up can be seen in Fig. 10.11.

Slowdown

The startup path modifier is used to change the printing parameters at the beginning of a path to ensure there are no issues with bead extrusion or adhesion during the transition to steady state. However, bead issues can also occur at the end of a path because the system must transition from steady state to shut off. The tip wipe modifier can be implemented to add additional pathing and ensure the nozzle gets wiped off, but a separate modifier can be used to alter the existing print path and modify the print parameters when approaching the end of the path. This path modifier is called a slowdown.

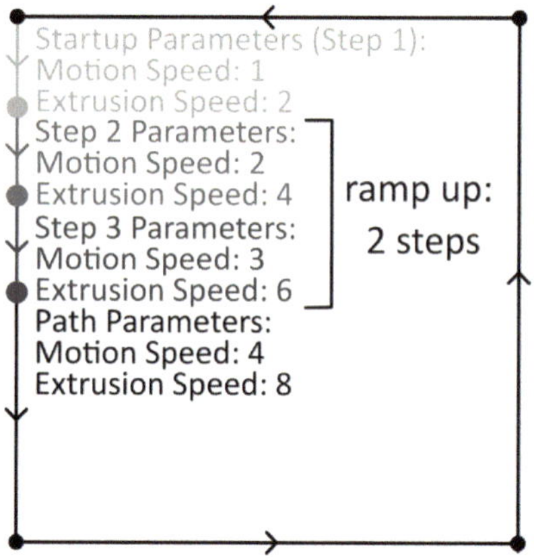

FIG. 10.11 All of the path segments in a startup path with a ramp up.

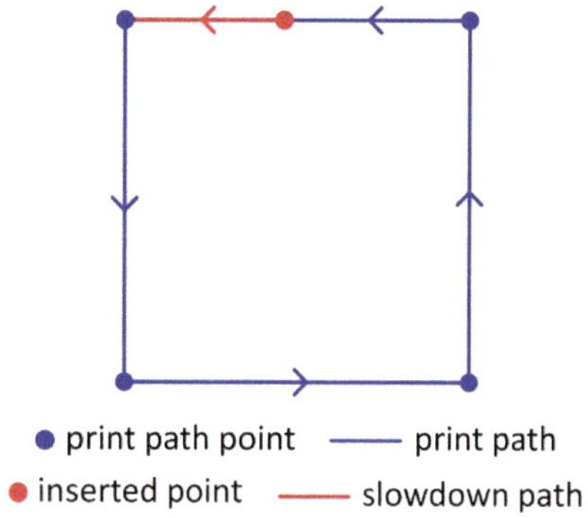

FIG. 10.12 A square toolpath with the slowdown path modifier applied.

A slowdown works similarly to the startup modifier where the user can define the length, extrusion speed, and motion speed of the slowdown. A ramp down functionality can also be implemented where the user can define a number of steps to transition from steady state to off. Fig. 10.12 shows a closed-loop toolpath with a slowdown modifier.

For some systems, the mismatch between motion deceleration and extruder deceleration is significant. These systems work best with a slowdown where the extruder speed is set to zero (off) before the end of the original print path, because the release of pressure in the extruder is sufficient to extrude enough material to complete the remaining path length. This is called "coasting"; the extruder is turned off, but the print path has not yet been completed.

Prestart

The startup modifier is used to handle the transition from start to steady state and can be applied to all path types. However, a similar path modifier, with a slightly different application, is the prestart. A prestart is only applicable to space-filling toolpaths that are bounded by an open- or closed-loop contour that has already been printed on the same layer. A prestart extends the space-filling toolpath by adding additional segments at the start of the path. These additional segments are usually very short and print over the bounding contour with a low extrusion rate, so that the space-filling toolpath's bead is bonded to the contour. Like a startup, this ensures the toolpath properly adheres and prints correctly. The difference is that a prestart traces over a path that has already been printed, which means the extrusion amount is low and can be completed faster. This ultimately saves time, compared to a startup modifier, and increases the stability of the print by bonding the infill and contour together. Fig. 10.13 shows an infill path before and after a prestart modifier has been added.

Prestart parameters include speed, length, and extrusion rate, which can be configured through user settings. Length is used to define the total length of the prestart path that will be added to the beginning of the space-filling toolpath, and this total length is divided between a contour traverse and a connecting move. A prestart path can be assigned its own speed and extrusion rate, independent of the space-filling toolpath. To path a prestart, the starting point of a space-filling toolpath is used to add a point at the closest location along the closest bounding contour path. A connecting path segment is added between the start point of the space-filling toolpath and the new point. Then, the prestart traverses the points of the contour until the desired prestart distance is achieved,

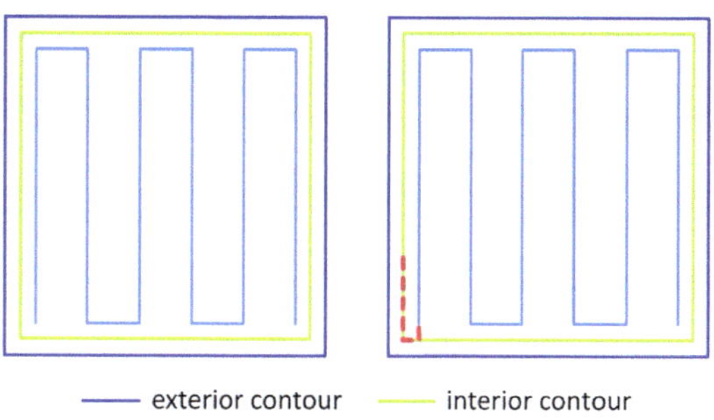

FIG. 10.13 Infill toolpath before and after the prestart modifier has been applied.

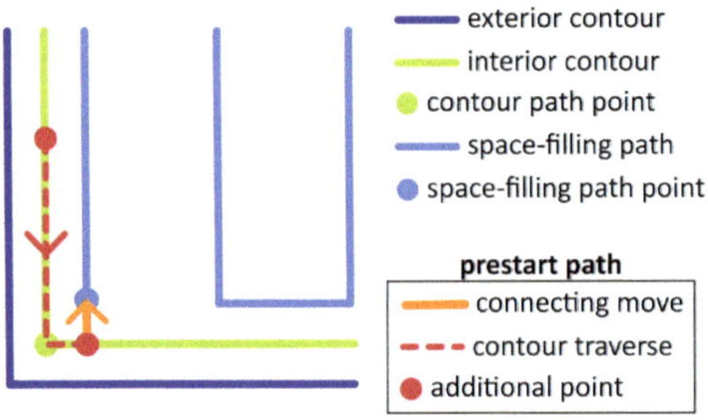

FIG. 10.14 Close up view of the prestart path modifier showing the connecting move.

adding additional points wherever needed. Fig. 10.14 shows a zoomed in view of the path segments in the prestart.

Conclusion

This chapter has outlined various path modifiers that add functionality to an existing toolpath, to help ensure a successful print: tip wipes, startups and slowdowns. Tip wipes are used to add pathing to the end of an existing toolpath, so that the nozzle can be wiped off after a motion is completed, and come in three varieties: forward, reverse, and angled. A startup is a modifier used to alter the beginning sequence of a path to make sure the bead extrudes and adheres properly, and to ensure there is a smooth transition from extruder off to a steady flow state. A slowdown, the counter to startup, is used at the end of a path to bring the extruder and printer motion to a smooth stop without excess material flow. A slowdown with the extruder disabled is called a coast. Lastly, the prestart modifier is used to add additional pathing to space-filling toolpaths so that the infill or skin pathing is properly adhered to the bounding contour. These modifiers may not be used on every print or every machine, but proper usage can be the difference between a successful versus failed print.

References

[1] P. Chesser, et al., Extrusion control for high quality printing on Big Area Additive Manufacturing (BAAM) systems, Addit. Manuf. 28 (2019) 445–455.

[2] M.S. Floater, Generalized barycentric coordinates and applications, Acta Numerica 24 (2015) 161–214.

CHAPTER

11

Travels, optimizations, and ordering

Michael Borish[a], Alex Roschli[a], Abby Barnes[a], Eric MacDonald[a,b], Peter Wang[a] and Thomas Feldhausen[a]

[a]Manufacturing Demonstration Facility, Oak Ridge National Laboratory, Knoxville, TN, United States, [b]University of Texas at El Paso, El Paso, TX, United States

Abstract

With all the necessary build pathing in place from the preceding steps, the final task is to organize the build pathing via directionality, ordering, and connectivity. Directionality controls closed-loop pathing by deciding whether construction occurs clockwise or counterclockwise. Ordering determines whether construction occurs outside-in or inside-out, by manipulating what path types will be printed first. Connectivity links all the pathing together, which is achieved through the insertion of travels. Connectivity has numerous approaches, but only the most common will be addressed in this chapter. In addition to these manipulations, other considerations, such as spiral pathing and unique cases for infill, will be discussed.

Introduction

With the major pathing computations complete, the final step of producing a complete pathing solution is what is referred to as "post-processing." In this stage, the construction paths produced in the previous steps are optimized, ordered, and connected according to the specific constraints that are of interest to the user. For each of these pathing manipulations, there are numerous approaches. This chapter will touch on common issues and solutions for each topic, but readers should be aware that this is by no means an exhaustive review.

https://doi.org/10.1016/B978-0-443-15286-3.00014-X

The first steps of post-processing are ordering and directionality, which are discussed in the next section. These manipulations should be performed before inserting travels or else the travels may have to be altered to account for the order and direction of paths. Once all paths have been ordered and, if appropriate, assigned a directionality, travels are inserted to efficiently connect all the paths together.

Directionality and ordering

Directionality refers to the direction of construction of closed-loop contours. Recall from earlier chapters that an STL does not have any specific ordering with respect to its faces, but once cross-sectioning has occurred, an ordering is typically imposed on the segments that comprise the resulting polygon. This is known as the winding order, which was included in the discussion of the erosion and dilation offsetting processes in *Chapter 6: Closed-loop toolpath generation.* Typically, points that represent solid geometry are oriented counterclockwise, while points that represent holes are oriented clockwise. This means that, as offsets are computed, each new offset will follow the same winding order, like the object shown in Fig. 11.1. The outer perimeters of this object are the result of offsetting a contour representing a solid body, while the inner perimeters are the result of offsetting the hole. As a result, the outer perimeters are ordered counterclockwise, while the inner perimeters are clockwise.

However, this directionality is not necessarily ideal for construction purposes because it adds unnecessary inconsistency. To increase consistency, all pathing of the same type is grouped in what is called a "region" and then identical manipulations are applied to all paths of a region. For example, all perimeters make up a region and would collectively be designated as either clockwise or counterclockwise, regardless of their original winding order. Similar logic can be applied to insets;

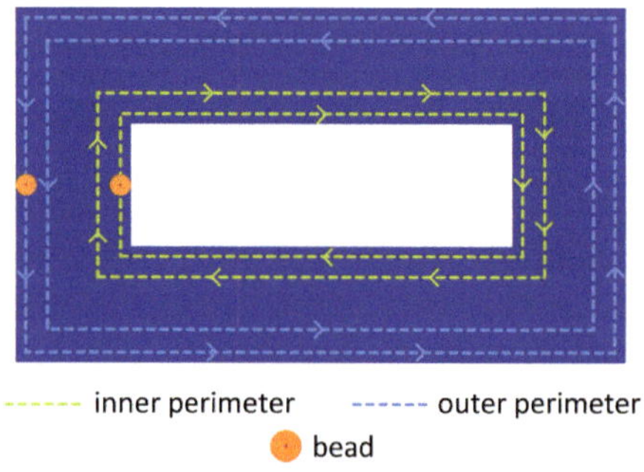

FIG. 11.1 Winding order for closed contours. Contours that arise as the result of solid boundary definitions will continue to be counterclockwise while those that arise from hole boundaries will continue to be clockwise.

because perimeters and insets represent closed-loop pathing, applying these concepts is straightforward. This logic can also be readily extended to concentric-style pathing for skins and infill. However, the directionality of other skin and infill patterns, as well as skeletons, is not as easily determined because those path types do not form closed loops. In those cases, the directionality of an open path is derived implicitly by the chosen strategy for travel insertion, which will be discussed in the next section.

Once directionality has been determined, there is one other unique optimization that can be applied: spiralization. Spiralization refers to the manipulation of paths into a spiral for continuous printing. The primary characteristic of a spiral path is a continuously increasing Z coordinate, such that layer transitions are unnecessary. Spiral pathing can be produced from a planar slicing approach through interpolation of the Z coordinate of each path. Take for example an object made up of a single circular perimeter, like the one shown in Fig. 11.2. Before spiralization, there would be several layers, each with a flat, circular perimeter. It is easy to interpolate Z coordinates by using the ratio of the length of the current path segment to the total path length. If a segment were to be 10% of the total length of the path, for example, then it would have its Z interpolated by 10% of the total layer height from start to finish. In this way, each segment will contribute a percentage of Z interpolation relative to its length and cover the entire layer height upon completion of the entire path. Once that interpolation is completed within each layer, a second interpolation between layers is necessary. This second interpolation connects the end of each layer's path to the start of the next layer's path, repeating as needed until the last layer is reached.

In this example, the interpolation for layer connectivity is a simple linear interpolation to the nearest neighbor. Despite its simplicity, this nearest-neighbor approach can work well even for geometries that change, if care is taken with the search. For example, instead of this cylindrical object, consider a cone. The diameter of each subsequent layer shrinks slightly, which means the points of the contours will not align as in the case of the cylinder. This makes it difficult to use a nearest-neighbor approach to connect contours to create a spiral. However, the amount of shrinkage is typically constrained by the bead width because consecutive

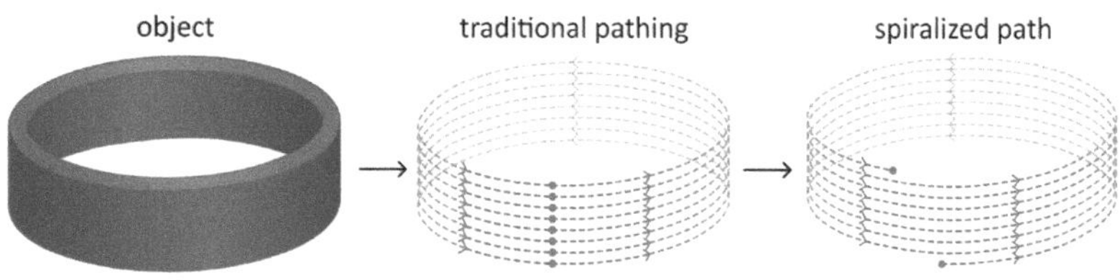

FIG. 11.2 Simple spiralization example. In the center, a traditional pathing approach shows several distinct start and stop points (solid dots). On the right side, the circular contours from each layer of the traditional pathing solution have been interpolated and connected to form a single, spiral path with one start point and one stop point.

layers cannot shrink more than one bead width without collapsing from a lack of support. The bead width is a known setting, so the inter-layer interpolation can use it to determine the closest contour on the next layer to connect the current spiral to. Ultimately, the interpolation and connectivity strategy will be dependent on a variety of concerns, including the pathing complexity, but this example shows an easy way to begin investigating potential approaches.

In addition to manipulating directionality, each region may also be ordered with respect to the centroid of the object. Remember that the two main categories of print paths are closed loops and open loops. Generally speaking, closed-loop paths are farther from the centroid of the object than open-loop paths. For example, perimeters and insets are always closed loops and are the farthest from the centroid. Skins and infill, however, can be closed or open loops and usually take up the space between the centroid and the perimeter of the object. Skeletons, on the other hand, are always open loops and usually fall on or near the object's centroid. Therefore, following the typical order of path generation (perimeters ->insets ->skins ->infill ->skeletons), layers are usually printed from the outside in. However, when organizing paths according to region, there is no reason that the order cannot be changed. It could be inverted, for example, to print a layer from the inside out.

Individual paths within a region can also be ordered, but with respect to the centroid of the region, as opposed to the entire object. This is straightforward for closed-loop paths, which each have a clear centroid, but is more difficult to apply to open-loop paths because they often do not have a clear centroid. For that reason, closed-loop paths are often considered better "defined" than open-loop paths. Additionally, there are an enormous number of methodologies for ordering open-loop paths, and it is hard to discern an optimal approach.

This is illustrated by Fig. 11.3. The shape shown in Fig. 11.3A is composed of mostly closed-loop toolpaths. There are only two skeletons, one near the center of the square and another in the small section of geometry in the top-right corner. In this case, an outward to inward ordering is straightforward, even for open-loop paths. However, in Fig. 11.3B, the ordering is less clear. In this truss example, skeleton pathing runs through

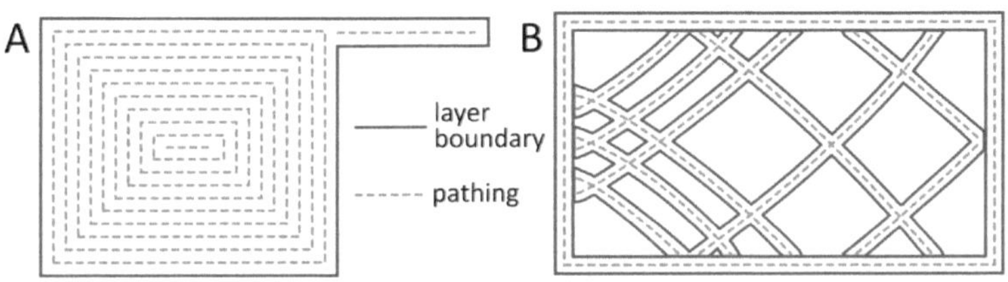

FIG. 11.3 In A, an outward to inward ordering is readily calculable even with open-loop pathing. In B, an outward to inward ordering is unclear.

most of the object and there is no clear ordering to be imposed. In such cases, the final choice of ordering is not based solely on the geometry in question, but also includes other variables, such as the process and material being used, as well as the required mechanical qualities of the final object.

Travels

Once directionality and ordering are determined, travel insertion can begin. Travels are one of the major computations done as part of post-processing and serve to impose connectivity on the rest of the pathing solution. Travels are any motion of the machine during which no material is extruded. During a travel, no construction is happening because no material is being introduced to the build. To maximize the efficiency of the printing process, in most cases, travels have their path length minimized and are performed at the maximum speed that the machine allows.

A travel path is typically broken into three distinct segments: a raise, a linear planar motion, and a lower. An example travel is shown in Fig. 11.4. When looking at the path of a travel, it's natural to ask, why bother having these distinct segments? If minimization of time is a priority, couldn't the end effector simply move from path to path, without the raise/lower? Alternatively, if the pathing must raise/lower, couldn't these motions be combined into a diagonal motion?

The answer to both these questions is yes. The pathing could forgo the raise/lower or combine them into a diagonal motion. The reason that neither is typically done is to guarantee collision avoidance. For the 3-axis, planar construction of a single object, requiring a travel to consist of these distinct motions ensures that the machine will never collide with the object. Naturally, guaranteed collision avoidance becomes more interesting when incorporating additional degrees of freedom during construction or slicing at off-angled planes or curves surfaces. Those concepts will be touched on in later chapters. For now, this chapter will focus on travels for a typical planar process that ascends the Z axis. Using the three distinct motions, more than one object can also be simultaneously

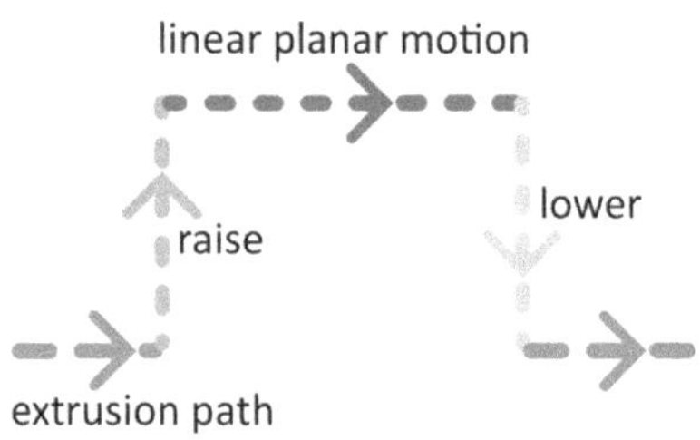

FIG. 11.4 Side view of travel motion. To move from one extrusion path to another, travels are typically composed of three distinct motions: a raise, the planar travel, and a lower.

constructed without fear of collision, as long as their bounding volumes do not intersect (though there are some extreme overhangs that need to be considered). The distance that the machine raises and lowers before performing the planar travel motion is specific to the machine in question but is usually larger than the layer height of the object under construction.

Travels can be incorporated into pathing solutions in a variety of ways. In general, strategies for travel insertion are distinguished by their approaches to three types of travels: those that move between layers, those that move between distinct polygons, and those that move within a polygon. These three types of travels will be referred to as layer, island, and path, respectively. Each will be defined and discussed in the following sections.

Layer travel insertion

Layer travel insertion is the first travel optimization to happen and is only applicable when constructing more than one object, because multiple layers only exist when simultaneously printing different objects. This type of travel optimization is likely to be the one that is least encountered when discussing "traditional" 3D printing. In fact, many slicers do not allow control over travel priority between layers. This is because the traditional 3D printing only builds one object at a time, or multiple copies of the same object with the same settings. When constructing only one object, there is only one set of layers, so the layers are evaluated sequentially, and no optimization is necessary. In the case of multiple identical objects, any selected order is usually sufficient, as long as it is maintained throughout the print.

However, the optimization of layer priority for travel insertion can be important when constructing more than one object in less traditional settings. Though providing this control increases the pathing complexity, it is necessary for more advanced pathing solutions. For example, if constructing two or more objects with different layer heights, collisions can occur when traveling from one object to another if the layers are not prioritized correctly. Additionally, layer priority is a significant issue when slicing objects at off angles. For example, building two objects at a 45-degree angle down the *X*-axis requires a specific layer ordering to prevent collisions. Fig. 11.5 illustrates these examples; Fig. 11.5A shows an example of two objects with different layer heights, while Fig. 11.5B shows a 45-degree slice down the *X*-axis. In both these cases, ordering of layers of the objects is necessary to avoid collisions and effectively construct the objects.

The method of choosing how to insert travels between layers is termed layer prioritization. There are three general approaches to layer prioritization, regardless of the angle of the cross-sectioning plane:

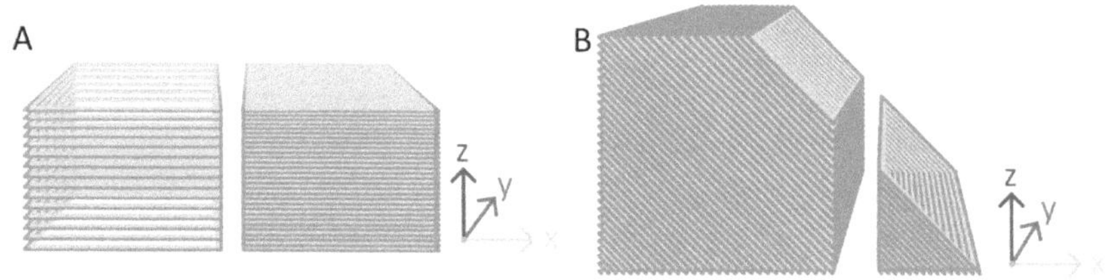

FIG. 11.5 Examples of layer ordering considerations. (A) Two objects with different layer heights. The object on the left has twice the layer height as the object on the right, and the layer ordering must account for the faster growth in height. (B) Slicing at a 45-degree angle. Layer ordering must account for the proximity of the layers of the two objects, such that no collisions occur.

- *By part.* Prioritize all layers that are contained in one object before moving to another. This strategy is useful for fully constructing an object before moving on to an identical copy but assumes sufficient empty build volume to avoid collisions.
- *By layer number.* Prioritize layers according to their number in each object. For example, layer 1 of object 1, then layer 1 of object 2, layer 2 of object 1, and so on. This strategy is useful for multiple objects with identical or nearly identical settings. For example, multiple objects with the same or nearly the same number of layers and equivalent layer heights.
- *By height.* Prioritize layers based on height from the bottom of the build volume. This strategy is useful when constructing multiple, heterogeneous objects or objects at angles other than horizontal. More detail about this strategy for off-axis printing will be discussed in *Chapter 15: Off-axis and nonplanar slicing.*

Using these strategies, the optimization algorithm can decide how to insert travels between layers and which layer to insert travels in first. Then, the algorithm moves on to evaluating travels within the layer, which includes both island and path travels. These are the typical travels seen in 3D printing and will be discussed together in the next section. Note that, at this point, no travels have actually been inserted. These strategies first function to locate where travels need to be inserted; insertion of the travels comes after the complete evaluation of layer, island, and path locations.

Island and path travel insertion

Recall from previous chapters that an island is any polygon representing a solid piece of geometry for which construction paths should be generated. Some slicers treat travels to an island and travels within an island as one in the same, while other slicers separate the two. The former has the benefit of making travel insertion easier at the cost of less user control, while the latter is more complicated but allows greater control over, and more flexibility in, the pathing solution.

If treated separately, once a layer is chosen by layer optimization, the islands that exist within that layer are evaluated for travel insertion based on various strategies. Some of the most common strategies for island travel insertion are:

- *Next closest.* The most common travel strategy. This strategy selects the next closest, unvisited island from the current position of the construction point.
- *Next farthest.* The opposite of the previous strategy. This strategy selects the farthest, unvisited island from the current position.
- *Shortest distance.* This strategy calculates the true shortest combined distance of all travels between islands. This will be discussed in more detail below.
- *Least recently visited.* This strategy selects the island that was visited most distantly in the past. For example, if you have a layer with three islands and visit them in order of 1 ->2 ->3, this pattern would be repeated for subsequent layers. Layers with new additional islands get added to the front of the list because they have not been visited. This makes the strategy functionally similar to a first-in first-out (FIFO) queue.
- *Random.* As the name implies, this strategy randomly selects the next island.
- *Consecutive.* A consecutive marching strategy that iterates over permutations of the island order. If for example you have three islands, the consecutive strategy would visit island 1 ->2 ->3, then 2 ->3 ->1, followed by 3 ->1 ->2, before repeating.
- *Custom point.* A user-defined point in the build volume from which travels should be calculated. This can be used in conjunction with the other strategies listed to provide an override location, instead of using the current position of the construction point. This is often used to guarantee an explicit and repeated pattern for every layer.

The "next closest" is the most common approach because it is a greedy algorithm, which is easy to compute and usually results in travels that are, for the most part, reasonably short. However, as object and machines scale in size, the need to optimize the length of travels may become necessary. For example, consider the difference in size between a desktop system and a large-format industrial machine, such as Big Area Additive Manufacturing (BAAM) [1]. A typical desktop system has a build volume of one to two feet for each axis, while BAAM's is 10–20 feet. At that size, nonoptimal travels can significantly increase build time because the machine may travel a large distance. In such cases, next closest may be insufficient and computing the true shortest distance is more appropriate.

Unfortunately, computing the shortest travels among all islands in a layer is a variation on the traveling salesman problem (TSP) [2]. TSP is a well-known computer science problem often phrased as the following: "Given a list of cities and the distances between each pair of cities, what is the shortest possible route that visits each city exactly once and returns to the origin city?". In this case, the cities can be represented by islands and the salesman by the build point or nozzle. Though we can relax the requirement about returning to the original city, the problem is still non-deterministic (N) polynomial time (P) hard, referred to as "NP-hard." This means that the problem is at least as hard as other NP problems and there is no known algorithm for computing the optimal solution beyond trying every combination. The number of combinations grows factorially, and so, the size of the problem quickly spirals out of control.

Despite this, there are several approaches that can be used to great effect to reduce overall travel time. If the number of islands is not too great, a graphics processing unit (GPU)-based solution can be used to directly compute the shortest path in a reasonable amount of time [3]. Alternatively, heuristic approaches are quite popular and can produce a solution to within a certain percentage of optimal. One of the most common heuristics is the Christofides algorithm [4]. The exact approach will vary by circumstance and readers are encouraged to investigate what approach works the best, considering the complexity of the geometry and machine in question.

Regardless of strategy, an island will be chosen for travel. Similarly to the layer travel optimization, at this point, no travels have been inserted. The next step is to choose a path within the particular island and, finally, insert the necessary travel. This inserted travel will move from the current position of the machine to the start of the selected path. Then, the position of the build location is updated, and the process continues, respectively moving through all paths, regions, islands, and layers. The process continues to repeat the relevant strategies as it moves through all the paths in a region, all the regions in an island, all the islands in a layer, and all the layers in the build.

The strategies for path travel optimization are unsurprisingly varied but share a significant overlap with the island optimization strategies that were previously listed. Using one of those same strategies, a specific path can be chosen, but the choice is also impacted by the region and path ordering described earlier in this chapter. Recall that it is possible to reorder regions and pathing within those regions. So, the next closest path in an island may not necessarily be the correct next path according to the order of regions or paths. Ultimately, the choice is a matter of bookkeeping as the appropriate order is layered upon the selected travel strategy.

One special case of travel insertion occurs for infill paths, which were discussed in *Chapter 7: Space-filling toolpath generation.* When infill paths are generated, only the lines from the projected infill pattern are computed (for reference, see Fig. 7.6). The small, connecting segments are created during the travel insertion stage, and they can either be build segments or travel segments, depending on whether a connecting segment would cross other paths or region boundaries. This determination cannot be made until travel insertion begins for the infill region. When a travel is being generated, if a particular travel segment does not cross another path or region, it is transformed into a small build path. Otherwise, a travel is inserted as normal. This ultimately gives rise to the raster behavior that is characteristic of the line infill pattern.

This approach requires less effort compared to potentially having to shuffle and reconnect infill segments if those small, connecting, raster lines had been created during generation of the original infill pattern. This consideration is, generally, only necessary for infill patterns that exhibit a raster-like behavior.

Conclusion

This chapter has introduced numerous manipulations of construction paths, which were discussed in the previous chapters. These include manipulation of path directionality, ordering, and travel insertion to impose connectivity. Unique manipulations were also explored, including spiralization, a process that reduces travels by interpolating the layer height across a path and connecting the end of a layer's path to the start of the next layer. The unique evaluation of travel insertion for raster-like infill patterns was also explained. Ultimately, the manipulations of this chapter are numerous and varied, and there is no one correct approach. However, the approaches presented here form the foundation upon which more advanced solutions may be built. The next chapters will focus on considerations for some specific advanced solutions.

References

[1] L. Love, et al., Commercialization of Big Area Additive Manufacturing, ORNL Report, *BT-, vol.* ORNL/TM-20, 2020.
[2] Traveling Salesman Problem, 2022. Wikipedia, Wikimedia Foundation https://en.wikipedia.org/wiki/Travelling_salesman_problem.
[3] M. Borish, C. Wade, A GPU-based approach for path planning optimization via travel length reduction, Proc. Manuf. 53 (2021) 310–317.
[4] Christofides Algorithm, Wikipedia, Wikimedia Foundation, 2022. https://en.wikipedia.org/wiki/Christofides_algorithm.

CHAPTER

12

Toolpath considerations for extrusion: Pellet, filament, concrete, and thermoset

Alex Roschli[a], *Michael Borish*[a], *Abby Barnes*[a], *Peter Wang*[a], *Thomas Feldhausen*[a], *and Eric MacDonald*[a,b]

[a]**Manufacturing Demonstration Facility, Oak Ridge National Laboratory, Knoxville, TN, United States,** [b]**University of Texas at El Paso, El Paso, TX, United States**

Abstract

There are a variety of extrusion and deposition-based processes for additive manufacturing, such as fused deposition modeling (FDM), fused filament fabrication (FFF), and directed energy deposition (DED). These processes can use a variety of materials, including thermoplastic pellets, thermoplastic filament, thermoset, concrete, metal wire, metal powder, and more. For all these processes and materials, the underlying toolpath strategies for 3D-printing an object are the same: the object to be printed is sliced into layers, then the layer is optimally fit with toolpaths of a given width, based on specific user settings. However, based on the specific process being used, additional modifiers and toolpathing strategies may be employed to maximize the capabilities of the process and successfully construct an object. This chapter explores the different pathing considerations and additional strategies to be employed in extrusion processes that use thermoplastics, thermosets, and concretes.

Introduction

The previous chapters of this book discuss the fundamental toolpathing concepts: slicing an object, fitting toolpaths, and applying path modifiers to have a successful print. These concepts are broadly applicable to most additive manufacturing (AM) processes and for a variety of machines, using a variety of materials. However, there exist additional pathing concepts and considerations that are more specialized and don't apply to all processes.

https://doi.org/10.1016/B978-0-443-15286-3.00004-7

This chapter will focus on toolpathing concepts that are specific to extrusion processes that utilize thermoplastic (both pellet and filament), thermoset, and concrete materials. Broadly speaking, the strategies that will be discussed involve modifications to toolpaths and the addition of features to compensate for machine limitations and the different flow characteristics of different materials.

Thermoplastic extrusion

Thermoplastic extrusion is the most popular form of AM due to the low cost of entry for machine ownership. Systems extruding thermoplastic materials are typically grouped into two major categories: pellet and filament. Pellet materials are low-cost and readily available because of the large quantities used in injection molding. Filament is more specialized for AM and is typically made from pellets. Pellet extruders are more complex than filament extruders but can be more easily scaled up to larger nozzle sizes and higher volumetric throughputs. For both systems, the pathing approaches are relatively similar. The differences stem from the control of the extruders.

Extruder control

Filament extruders are positive displacement pumps that use a gear, or set of gears, to grab and push filament through a heated tube and nozzle, called a hot end [1]. The volume of material output can be directly correlated to the volume of material input. To calculate the command for filament input to the extruder, Eq. (12.1) is used. Eq. (12.1), where F_D is filament diameter and N_D is nozzle diameter, can be used to relate the input filament length (F_L) to the printing path length (P_L).

$$F_L * \pi * \left(\frac{F_D}{2}\right)^2 = P_L * \pi * \left(\frac{N_D}{2}\right)^2 \quad (12.1)$$

Eq. (12.1) can be simplified to solve for the input filament length (F_L), given the total print path length (P_L), using Eq. (12.2). The F_L value can be used for positional control by commanding how much filament to extrude for the path.

$$F_L = \frac{P_L * \left(\frac{N_D}{2}\right)^2}{\left(\frac{F_D}{2}\right)^2} \quad (12.2)$$

Pellet extruders, on the other hand, are nonpositive displacement [2] and therefore cannot use the same method of positional control because there is not a defined distance or volume to feed into the extruder for the print path. They work through a screw-based auger that feeds pellets through a heated barrel and out a nozzle. The pellets melt from the barrel heat and the shearing action of the extrusion screw [3]. Gravity is often the force used to feed pellets into the extruder, but this can produce inconsistencies in the feed rate because the flow of pellets cannot be precisely controlled. This inconsistent feed causes dead space and air pockets within the extruder. To combat this issue, pellet extruders are commanded with velocity control, or more specifically, RPM (revolutions per minute) control, because RPM control can be modulated to help overcome the inconsistent input-to-output relationship. For example, although an extrusion RPM is commanded for the extruder motor at the start of the path and is used for the entire length of the path, it can be modulated for corners and other instances of acceleration/deceleration. For industrial systems with a programmable logic controller (PLC), the RPM modulation can be done in real time by the 3D printer based on what is currently happening along the path. For simpler systems, including those controlled by microcontrollers, the modulation needs to be done through the slicing process and the updated commands must be written to the g-code.

One such modulation is corner ramping. Corner ramping needs to be done because fast changes in direction often cause printers to briefly pause before the direction change, which leads to an excessive amount of material being deposited at the point of the direction change, causing "round" corners. To help compensate, the speed of the extruder can be gradually slowed while approaching a corner through a process called a "ramp down," and then gradually raised after the corner through a process called a "ramp up." The user can define the target motion speeds for the ramp downs and ramp ups, as well as the ramp up and ramp down distances, which are the distances from either end of the corner. The extrusion speed can also be similarly lowered and raised, which also helps combat the deposition of excess material.

Corner ramping in the slicing process works by analyzing paths to detect corners, defined here as direction changes that exceed a user-specified threshold, and modifying the path segments before and after the direction change. To do this, a corner or direction change is first found by finding the angle of each path segment with respect to the following segment to see if the angle exceeds the user-defined threshold for corner ramping. Once a corner that meets the threshold has been found, the path segment(s) on either side of the corner can be modified based on the user settings. To appropriately apply the path modifications, the path segments must be broken into smaller segments using Eqs. (12.3)–(12.5), which are the same as Eqs. (10.1)–(10.3) in *Chapter 10: Path modifiers*.

$$M = 1 - \left(\frac{\text{Ramping distance}}{\text{Segment length}}\right) \tag{12.3}$$

$$P_X = (1 - M) * P_{1X} + M * P_{2X} \tag{12.4}$$

$$P_Y = (1 - M) * P_{1Y} + M * P_{2Y} \tag{12.5}$$

In Eq. (12.3), the segment length refers to the full length of the segment being modified for ramping, while the ramping distance is the length of the ramping segment to be added, and M is the resultant modifier to be used for Eqs. (12.4) and (12.5). In Eqs. (12.4) and (12.5), P_X and P_Y are the X and Y components of the point where the original path segment is broken. P_{1X} and P_{1Y} are the corresponding components of the start point of the original path segment and P_{2X} and P_{2Y} are likewise the components of the end point of the original segment. Fig. 12.1 shows the pathing for a square before and after corner ramping is applied.

Filament extruders, because they typically work through a gear pushing on a filament rod, are more easily able to accelerate and decelerate than a pellet extruder, which has the thermal mass of molten polymer surrounding the extrusion screw. The inability to accelerate and decelerate quickly on a pellet extruder can cause bead width inconsistencies on corners [4]. This means that corners, and other scenarios where the machine is changing speed, tend to have a more consistent bead width on filament extruders. It also explains why filament extruders are less likely to employ a corner ramping functionality.

Another benefit to filament extruders is the ability to pull back on the filament, called retraction. This is particularly helpful for the end of the path where the flow needs to be stopped quickly. This retraction move for filament extruders requires additional control commands to pull back on the filament by a known amount, at the completion of a path. At the start of the next path, the distance of the retraction should be added as a priming extrusion move to reload the filament into the nozzle and ensure extrusion starts properly. This priming motion would happen only at the extrusion axis, without moving the motion system.

Pellet extruders don't have the ability to retract because of how the screw-based auger system works. Simply turning the screw backwards

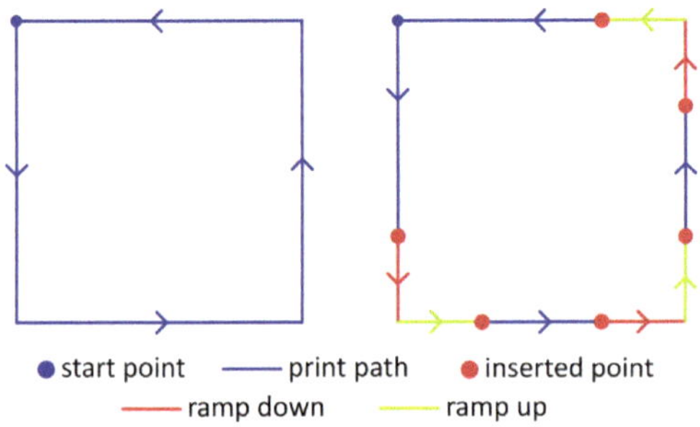

FIG. 12.1 Corner ramping applied to a square toolpath.

doesn't effectively withdraw material from the nozzle because the system is nonpositive displacement. There's also the concern that running the extrusion screw in reverse could cause some melted material to be retracted to the upper portion of the barrel where melting should not occur. This can result in an effect called bridging, where melted material forms a bridge over the screw such that it cannot be caught and pushed out of the nozzle by the turning of the screw. For these reasons, pellet thermoplastic extrusion systems cannot use retraction and must rely on path modifiers such as start-up, slow down, prestart, and tip wipe to assist with the starting and stopping of paths.

Managing temperature

Thermoplastic extrusion is a thermal process wherein the thermoplastic material is heated to the point of melting so that it can be extruded. Once extruded, the bead of thermoplastic material begins cooling down to ambient temperature. The rate of cooling is dependent on the material, temperature gradient, surface area to volume ratio, and a few other factors. The surface area-to-volume ratio, which is defined by slicing parameters, can be a large contributing factor for cooling, and this ratio can vary significantly between filament and pellets.

The surface area and volume of a bead are directly proportional to the size of the nozzle being used to print the bead. The standard size of a bead, if printed from a circular nozzle orifice, has a width that is 110% of the nozzle diameter (1.1*ND*) and a height equal to half that of the nozzle diameter (0.5*ND*). The extruded bead leaves the nozzle with a circular shape, but the printing process of extruding the bead directly onto a surface while moving the extruder horizontally creates a rounded rectangular shape. The flat top edge is created by the edge of the nozzle dragging across the top of the bead and the flat bottom comes from the flat surface on which the bead is deposited. A representation of a cross-section of this bead geometry can be seen in Fig. 12.2.

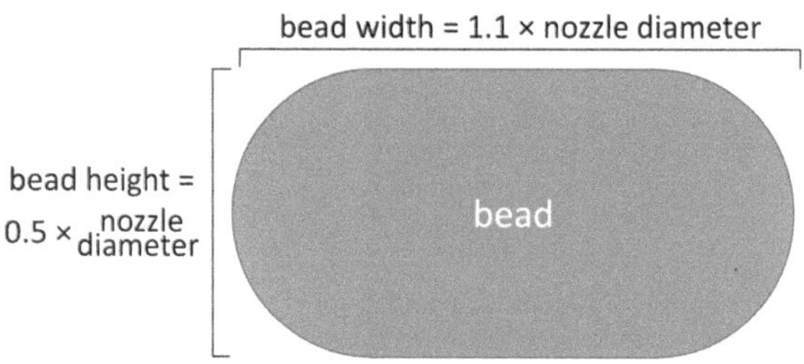

FIG. 12.2 Cross section of a bead with standard aspect ratio.

To approximate the surface area (*SA*) of a printed bead, the cross-sectional perimeter of the bead is multiplied by the length (*L*) of the bead. For this calculation, it is easiest to view the bead cross-section as a rectangle with half circles on each end, as seen in Fig. 12.3. The width (*W*) and height (*H*) of the rectangle are used, as shown in Eq. (12.6), to calculate the surface area. *H* is equivalent to both the height of the bead and the diameter of the two half-circle portions. The height of the bead can therefore be used to calculate the perimeter of the two half circles, which together make one whole circle. The remaining portion of the bead's perimeter is then the top and bottom edges of the rectangle portion, which is equal to two times *W*. *W* is equal to the bead width minus the two radii of the half circles – or, in other words, the diameter of the whole circle. As previously stated, the diameter of the circle is equal to *H*, so *W* is equal to the bead width minus *H*. All of these terms and their relationship to each other are illustrated in Fig. 12.3.

$$\mathrm{SA} = (2 * W + \pi * H) * L \tag{12.6}$$

The same bead's *SA* can be calculated by replacing *W* and *H* with the corresponding relationship to the nozzle diameter (*ND*). That is, *W* is replaced with 1.1*ND* – 0.5*ND* and *H* replaced with 0.5*ND*, as shown in Eq. (12.7), which can be simplified as Eq. (12.8). The bead surface area can then be found in terms of the nozzle diameter and total length, which indicates how the total surface area grows based on the nozzle diameter.

$$\mathrm{SA} = (2 * (1.1 * ND - 0.5 * ND) + \pi * (0.5 * ND)) * L \tag{12.7}$$

$$\mathrm{SA} = \left(1.2 + \frac{\pi}{2}\right) * ND * L \tag{12.8}$$

Using the same *W*, *H*, and *L* terms, the volume (*V*) of a bead can be approximated by multiplying its cross-sectional area by its length, as shown in Eq. (12.9). The volume of the bead can be found in terms *ND*

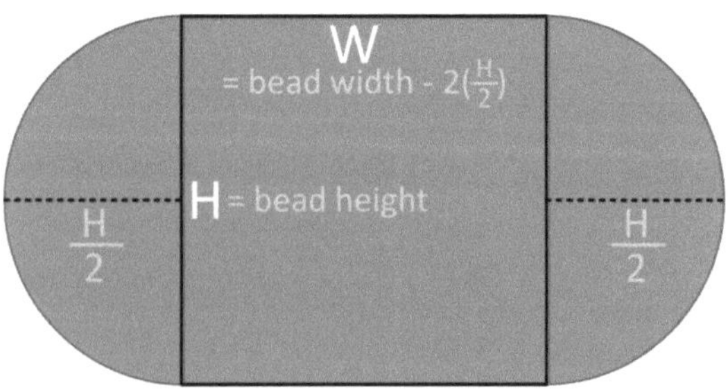

FIG. 12.3 An illustration of the terms used to calculate the surface area and volume of a bead (*W* and *H*) as well as their relationship to the bead width and bead height.

by replacing *W* and *H*, just as was done above for the *SA*, as shown in Eq. (12.10), which can then be simplified as Eq. (12.11).

$$V = \left(W * H + \pi * \left(\frac{H}{2}\right)^2\right) * L \tag{12.9}$$

$$V = \left((1.1 * ND - 0.5 * ND) * (0.5 * ND) + \pi * \left(\frac{(0.5 * ND)}{2}\right)^2\right) * L \tag{12.10}$$

$$V = \left(0.3 * ND^2 + \frac{\pi}{16} * ND^2\right) * L \tag{12.11}$$

Using the simplified equations, (Eqs. 12.8 and 12.11), the surface area and volume can be found in terms of the ND. As indicated by the ND^2 term in the volume equation, the volume grows much more quickly than the surface area with a growing ND. Ultimately, this means that larger nozzle diameters, which are seen on pellet extruders, have a smaller surface area-to-volume ratio, which causes a slower cooling rate. These larger pellet-extruded beads cool more slowly and thus need longer layer times to allow the object to adequately cool before printing the next layer.

The slicing process can account for the necessary cooling in two different ways. The first, and more common, method is feedrate scaling, which adjusts the feedrate, also known as print speed, to meet a desired layer time. The feedrate scaling factor is found from the ratio of the current layer time to the desired layer time, seen in Eq. (12.12). The current layer time is found by approximating the total time to print the layer by adding up the print time for each path segment of all the toolpaths in the layer. The desired layer time is a user-defined setting.

$$\text{Scaling factor} = \frac{\text{Current layer time}}{\text{Desired layer time}} \tag{12.12}$$

For forcing a minimum layer time, Eq. (12.12) is only implemented when the current layer time is less than the desired layer time. This produces a feedrate scaling factor less than one. If the current layer time is longer than the desired layer time, the resultant scaling factor is larger than one. When applying this larger-than-one scaling factor, the print feedrate would be increased to meet the desired layer time. This can be used on systems that want to force an exact layer time for every layer.

The second approach to additional cooling is fan control. Fans are common on desktop filament printers because of the effectiveness of the fan for quickly cooling a bead. The slicing software can turn the fan on/off and adjust the speed based on the toolpath and layer time.

Thermoset extrusion

Thermoset extrusion AM has a lot of similarities with thermoplastic AM extrusion, but with a few key differences. The thermoset material is often a liquid resin being extruded by a pumping system. This resin can have oozing and drooling issues like those seen in thermoplastic systems, which must be accounted for through toolpath modifiers such as tip wipes and start-ups/slowdowns. Thermoset, like pellet thermoplastic, doesn't have the ability to retract, but the nozzle can be closed to help stop flow.

Extruded thermoset material is a liquid that immediately begins curing into its final form through a chemical reaction. During the curing, the resin rapidly increases viscosity until it becomes a solid. However, at the initial moment of extrusion, the resin is a low viscosity material that behaves as a liquid [5]. This means that thermoset deposition has almost no ability to print over a gap between previously extruded regions, often termed a "bridge." Where this becomes an issue is infill structures, such as a rotating rectilinear infill, where the raster pattern rotates 90 degrees and must be able to bridge the gap between infill beads on the previous layer (Fig. 12.4).

The inability to bridge during infill makes it difficult to do sparse infill with a solid skin top layer because the skin fill material will just fall into the void space between previous infill beads, causing the top surface to print unevenly. This can be overcome by implementing gradual infill, which was discussed in *Chapter 7: Space-filling toolpath generation*. Gradual infill is typically implemented using the grid fill pattern so that bridging support is provided in two different directions, thus minimizing the bridging distance. The fill density of gradual infill is set to slowly increase from the defined sparse infill density to the 100% dense infill of the top skin. The increment for the density change is defined by the number of steps, or number of layers, of gradual infill. For example, a sparse infill density of 50% with three steps of gradual infill would result in three layers of gradual infill with the following densities: 62.5%, 75%, and 87.5%.

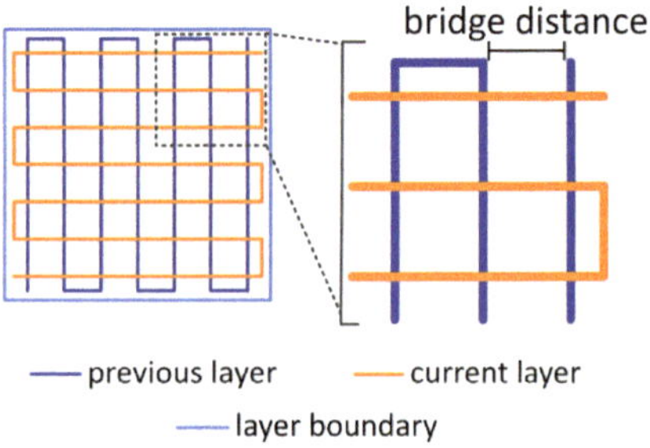

FIG. 12.4 Bridge distance as it applies to infill toolpathing.

Concrete extrusion

Concrete extrusion AM has a lot of similarities with thermoset extrusion AM. Like thermoset, concrete has no ability to retract material, and the material during extrusion should be considered a liquid. The extruded concrete material is curing and slowly becoming more solid to support the previous layer. This means precise control of layer time is needed to ensure enough time for concrete curing.

Traditional concrete pouring and manufacturing involves troweling the surface to make it smooth and shape it as desired. Because extruded concrete material has a low viscosity, it can be shaped by the nozzle to the desired geometry. This shaping is implemented via a trailing nozzle that extrudes material behind the motion of the extruder [6]. The nozzle orifice can be shaped to the desired output bead geometry, to act like a trowel for the concrete, like the rotary trailing nozzle shown in Fig. 12.5. A system with a nozzle like this requires two additional toolpathing controls that aren't used by systems with a traditional, circular nozzle orifice that extrudes directly down. The first is rotational control of the nozzle, acting as an additional motion axis for the system. The angle of rotation is calculated from the trajectory vector of a path segment and its preceding path segment. The rotation move should not be implemented by the machine controller as a coordinated move. Instead, it should be commanded immediately at the start of each segment, to ensure the nozzle is steered in the correct direction for the duration of the path.

The second change needed for this new nozzle setup impacts the end of an extrusion path. Because a circular nozzle stays above the top of the extruded bead, there is never a risk of colliding with extrusion from the current layer. However, a trailing nozzle cannot stay above the current layer; the nozzle must be within the volume of the layer to lead the bead

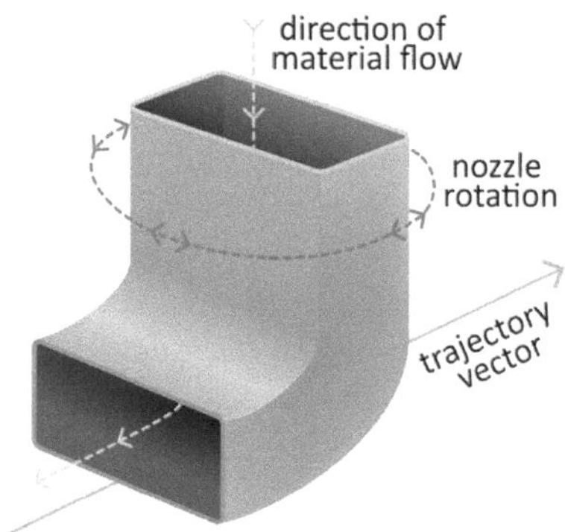

FIG. 12.5 A diagram of the motion of a trailing nozzle extruder.

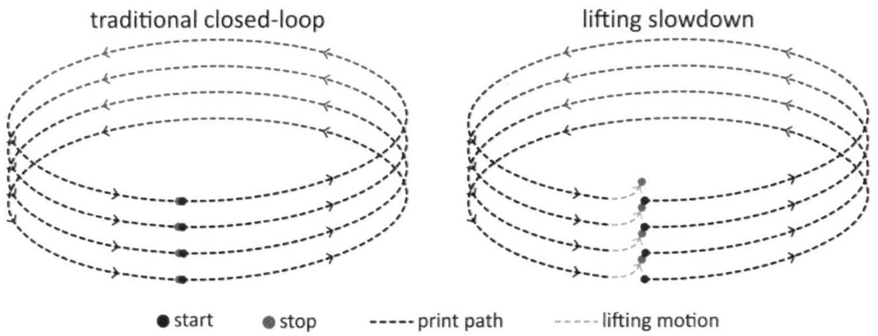

FIG. 12.6 Traditional circular pathing for four layers versus the same pathing with a lifting motion added at the end of the path during slow down.

as it is being deposited. This prevents a closed-loop path from printing using the traditional parameters, because the nozzle would collide with the beginning of the loop when attempting to closing the loop. To get around this, a lifting motion must be added at the end of a closed-loop path. This modification can be added to the slow down pathing so that the nozzle lifts by one full layer height as the slow down segment is completed. Fig. 12.6 shows a traditional closed-loop path trajectory and a closed-loop path trajectory with lifting slow down.

Conclusion

This chapter has outlined the considerations and modifications needed when toolpathing for extrusion-based additive manufacturing, including thermoplastic filament and pellet extrusion, thermoset epoxy extrusion, and concrete extrusion. Thermoplastic processes involve thermal extrusion, so heat and layer time must be managed by modulating print speeds and/or using cooling systems. Filament systems offer retraction and precise extrusion control and therefore can operate via positional extrusion commands. Pellet systems have less precise extruder dynamics without the option for retraction, so velocity control must be implemented for the extruder. Corner ramping can be added to paths to help control the flow on pellet extruders, and the same approach can be applied to thermosets and concretes. Thermosets have the added difficulty of a low viscosity extrudate, so pathing solutions, such as gradual infill, must be implemented to overcome the inability to bridge. Concrete systems often implement a nozzle geometry that extrudes material behind the motion, rather than from above as with thermoplastic and thermoset, to enable troweling or shaping of the extrudate. This nozzle configuration must be accounted for by controlling the angle of the nozzle and lifting at the end of the path to avoid collisions. Thermoplastic, thermoset, and concrete toolpathing solutions all rely on the same fundamental algorithms, and successful printing would not be possible without these additional modifications, making them essential to the capabilities of those processes.

References

[1] S. Park, F. Kun Kelvin, Polymer-based filament feedstock for additive manufacturing, Comp. Sci. Technol. 213 (2021) 108876.

[2] P. Chesser, et al., Extrusion control for high quality printing on Big Area Additive Manufacturing (BAAM) systems, Addit. Manuf. 28 (2019) 445–455.

[3] A.A. Hassen, et al., Additive manufacturing of composite tooling using high temperature thermoplastic materials, in: SAMPE Conference Proceedings, Long Beach, CA, 2016.

[4] A.C. Roschli, Dynamic Extruder Control for Polymer Printing in Big Area Additive Manufacturing, 2016.

[5] C. Hershey, et al., Large-Scale Reactive Extrusion Deposition of Sparse Infill Structures with Solid Perimeters, Oak Ridge National Lab.(ORNL), Oak Ridge, TN (United States), 2019.

[6] C. Atkins, et al., Construction-scale concrete additive manufacturing and its application in infrastructure energy storage, in: ASME International Mechanical Engineering Congress and Exposition. Vol. 84485. American Society of Mechanical Engineers, 2020.

CHAPTER

13

Toolpath considerations for DED: Arc and laser welding

Michael Borish[a], *Alex Roschli*[a], *Abby Barnes*[a], *Eric MacDonald*[a,b], *Peter Wang*[a], *and Thomas Feldhausen*[a]

[a]**Manufacturing Demonstration Facility, Oak Ridge National Laboratory, Knoxville, TN, United States,** [b]**University of Texas at El Paso, El Paso, TX, United States**

Abstract

Directed energy deposition (DED) is a form of additive manufacturing that uses localized melted metal, similar to traditional arc welding, to construct an object. In comparison to polymer systems, metal systems can exert additional control over the build process through manipulation of the pool of melted metal, called the "melt pool." This is especially important to maintain appropriate layer heights and create high quality builds. Additionally, metal systems allow site-specific control of the printed part, which enables advanced construction techniques that would otherwise not be possible. However, both of these types of control require support from the slicing software and, potentially, sensor feedback. This chapter will discuss these control processes and their basic implementation in a slicing program. The processes described here lay the foundation for more advanced solutions, particularly sensor feedback, that will be discussed later.

Introduction

In comparison to *Chapter 12: Pathing considerations for extrusion: Pellet, filament, concrete, and thermoset,* which focused on the specifics of polymer and composite-based systems, this chapter will focus on the specifics of metal-based systems, sometimes called directed energy deposition (DED) systems. In general, there are three types of DED processes: arc, laser, and electron beam welding. These three types of welding differ in the type of energy used to melt the metal but are similar in their pathing

https://doi.org/10.1016/B978-0-443-15286-3.00005-9

setups and processes. Both types use a robotic arm system instead of a gantry, which allows additional degrees of freedom for construction. However, the considerations for additional degrees of freedom will be covered in *Chapter 16: Pathing considerations for 5-axis systems*. This chapter will only focus on the slicing considerations that are necessary when working with metals.

In DED systems, either a metal wire filament or a metal powder is fed to the build point, where the energy source is directed and, in turn, where the metal is melted. The energy source is either an electric arc, an industrial laser, or a highly charged stream of electrons, depending on if the system is an arc, laser or electron beam system, respectively. At the build point, the metal feed that is melted by these energy sources is called a melt pool. To construct an object, the melt pool is continuously deposited on a metal build plate, or on top of previously deposited layers, and allowed to cool. Typically, the objects created from such processes are low resolution, with imprecise dimensions and features, and are subsequently machined to create the final part.

The behavior of the material in metal-based systems is significantly different than that of polymer-based systems. The viscosity of melted polymers is quite high, so the extruded material tends to maintain its shape. In comparison, melted metal is essentially a liquid, which introduces challenges when attempting to create high quality parts. There are a number of variables that affect the quality of metal-based AM, and a complete summary of them all is beyond the scope of this book. Instead, this chapter will discuss the ones that are readily supportable with modifications to the slicing process: height control, site-specific control, and an advanced construction process based on site-specific control.

Height control

Height control is fundamental to metal processes. Construction without height control is difficult because layers do not maintain a consistent height, due to the liquid melt pool. An example of this is shown in Fig. 13.1, in which the object in front was printed without height control and the object behind it was printed using height control. The difference between construction with and without height control is significant; the downward slope of the front object is a result of a lack of height control. Ultimately, such a build will fail, as the actual height deviates from the expected height more and more with each successive layer.

In order to implement height control, it's necessary to get consistent feedback about the current height of the build from the 3D printer. With relatively straightforward modifications, the slicer program can help gather this feedback, and the control system of the machine can then

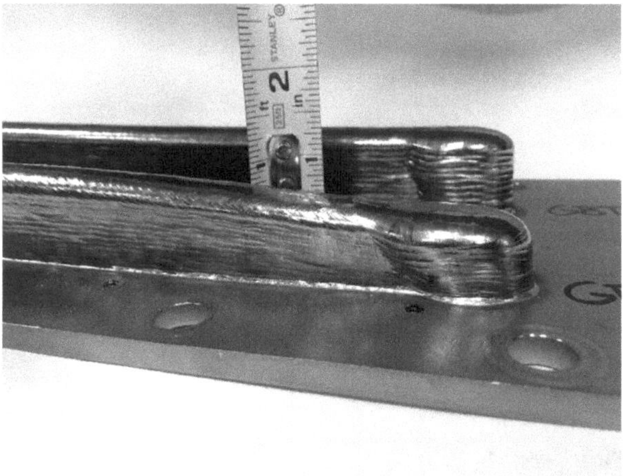

FIG. 13.1 Large format laser DED with (background) and without height control (foreground). Without height control, it is very difficult to successfully build an object due to the sloping behavior.

use feedback to alter the print parameters in real time, adjusting as necessary to compensate for changes in height [1]. The feedback required for height control can be collected in several different ways, but this discussion will focus on the use of a laser profilometer to measure the build height.

A laser profilometer is a laser displacement sensor and, as such, emits light onto an object and uses the light that is reflected off the object to measure the distance and angle from the laser to the object's surface. The distance of an object from a laser displacement sensor negatively correlates with the object's height, so a laser displacement sensor can collect height information. A laser profilometer is distinguished from other laser displacement sensors in that its laser is a line, not a single point. To gather the necessary information for height control, the laser profilometer scans a layer immediately after it has been printed. From the collected data, a height map is constructed. The height map is then passed to the control system to adjust the process parameters of the next layer, and the process is then repeated for each additional layer.

As far as pathing computation is concerned, the process of generating print toolpaths for DED systems is the same as what has been described in this book so far. However, in addition to these print paths, scan pathing also needs to be generated. Scan pathing is straightforward for a planar slice. Once a layer has been constructed, the laser profilometer will need to pass above the entire area of the layer to collect information. The most efficient way to cover the entire area is with a raster pattern. Computation of this raster pathing is based on information about both the object and the

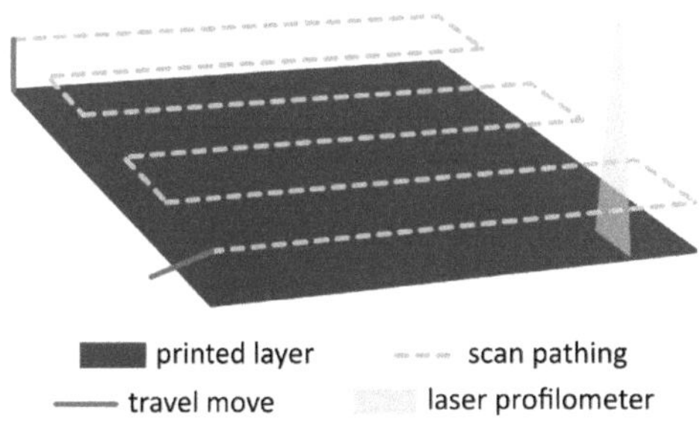

FIG. 13.2 Raster pattern for laser profilometer inspection of layer.

profilometer. The laser profilometer is treated like an extruder—the length of the line of the beam is akin to the bead width. Paths are placed such that the midline of the path correlates to the midline of the beam, and the bounding box of the object is divided by the beam width to find the necessary number of scan paths. An example of a raster scan path for a laser profilometer is shown in Fig. 13.2.

Naturally, spatial offsets must be accounted for because the center of the profilometer is not at the same location as the center of the extruder. The extruder needs to be moved out of the way, and the sensor must be moved into position. The locations required for those movements are input into the slicing software, like any other setting. The scan pathing, once computed, can be included as part of the g-code file with additional commands to allow the system to control the sensors.

After a layer has been scanned, the next step is to compute height maps, which requires various information that ultimately gets stored outside of the g-code file, in companion files. These companion files are usually generated by the slicer as part of the slicing process. Everything necessary to generate and analyze a height map is contained in the files, including information such as the polygon representation of the geometry to be constructed, its bounding box, expected layer heights, and physical parameters of the profilometer such as beam resolution and step distance between successive scans. Both these companion files and the g-code file should be supplied to the 3D printer for construction.

As the DED construction process with height control continues, it alternates between building and scanning. In general, a layer is constructed and then scanned using a raster pattern. The sensor data is then combined with the information from the companion files to produce an appropriate height map. This height map is then passed to the control system of the 3D printer, which compares the height map to the expected height. The control system then uses the comparison to decide how to manipulate various process parameters during the deposition of subsequent layers. For example, if the control system finds the height is too low, such as with the sloping effect shown in Fig. 13.1, filament input could be increased on the subsequent layer to place additional material in the appropriate areas. If the height was found to be too high, filament input could be reduced. Because of the complex nature of the metal melt pool, changes to rate of motion or laser power can also adjust the rate of deposition and, likewise, can be used to control the layer height. However, all these adjustments rely on data about the expected construction and on the data the laser sensor has collected, both of which depend on the slicer's generation of companion files and scan pathing.

For an initial setup of this process, the software that is collecting and interpreting the sensor results does not have to be contained within the slicer. However, if it is not, the slicer will probably need network connectivity to provide the sensing software with information about the profilometer and the expected build. This creates an open-loop configuration that is common among many current approaches to feedback in additive manufacturing. Executing the scan pathing generates sensor feedback, which flows to a separate software package. That software processes the feedback alongside information from the slicer and then informs the control system, which finally implements changes in print parameters. This setup is natural for an initial implementation, but it has limitations. Chief among those limitations is that the pathing solution that starts the process, generated by the slicer, is not involved in the later steps of the process and cannot provide true path replanning. True path replanning would require changing the fundamental paradigm of the slicer and is a topic discussed in detail in *Chapter 19: Closing the loop*.

Site-specific control

As previously mentioned, changes to the filament flow, rate of motion, or laser power can all affect the final build due to melt pool dynamics. That is an understatement, and the topic of melt pool manipulation is very large—so large that it cannot be adequately covered in a single section. That said, this section will detail one slicer-supported approach that allows manipulation of print parameters across the *XY* plane. This *XY* manipulation is called site-specific control and complements the manipulation of print parameters that affect the *Z* axis, which is provided by the height control described in the previous section [1–3].

The need for *XY* manipulation is illustrated by the simple cross-shaped object in Fig. 13.3. The print pathing for this object, shown in Fig. 13.3A, is easily computed as a single, closed-loop path around the outer edge, but a problem arises during construction. In the center of the object, the pathing does not move close enough to the pathing on the opposite side of the

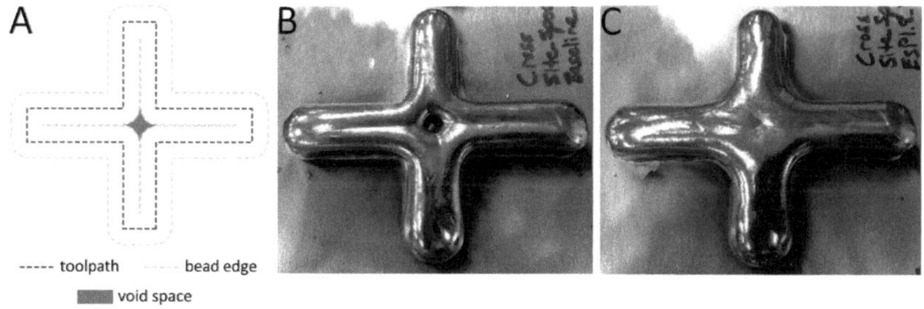

FIG. 13.3 (A) The toolpath for a cross-shaped geometry, note the void in the center; (B) object printed without site-specific control; (C) object printed with site-specific control. The additional control afforded through feedback allows successful construction of objects that would otherwise fail. *Credit: B.T. Gibson, Y.K. Bandari, B.S. Richardson, W.C. Henry, E.J. Vetland, T.W. Sundermann, L.J. Love, Melt pool size control through multiple closed-loop modalities in laser-wire directed energy deposition of Ti-6Al-4V, Addit. Manuf. 32 (2020) 100993.*

cross, leaving a hole in the middle. Under normal operation, with the expected bead width, the melt pool cannot cover this void (Fig. 13.3B). Even though the pathing is computed appropriately, the physical characteristics of the melt pool cause the build to fail. The hole could be filled with a skeleton, but the characteristics of the melt pool offer a solution that does not require the computation of additional paths or travels. By using site-specific control to adjust the print parameters near the center of the object, the melt pool can fill the void, as shown in Fig. 13.3C.

The slicer can implement site-specific control by generating a new type of mesh: the settings mesh. Usually, a slicer provides a build mesh that represents the expected geometry of the constructed object and each constructed layer. This mesh has pathing fit to it, as described in the previous chapters, and is what is passed along as a comparison for the height maps mentioned in the previous section. A settings mesh, however, does not have any pathing. Instead, this mesh is used to volumetrically control various settings in specific areas of the printer's build space.

Obviously, this volumetric control requires the ability to freely manipulate slicer settings. Many slicers do not allow free manipulation of settings. For example, the layer height is often set as a global variable, meaning that multiple objects can be loaded but they will all share a single layer height. This approach is too limiting when trying to exercise control over the build process. There is no reason that multiple objects could not have multiple layer heights. In fact, a single object could even have layer heights that change regularly throughout the build. Settings meshes allow a finer level of control than even object- or layer-specific settings, because they allow control over specific areas of objects and their layers. A slicer that allows this level of setting manipulation is required for implementing site-specific control.

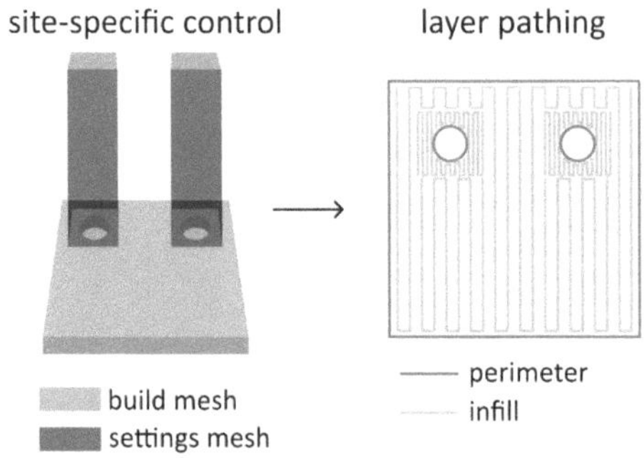

FIG. 13.4 Simple example of site-specific control using settings meshes. Pathing within the settings mesh is densified compared to the remaining volume.

An example of a build mesh with settings meshes is shown in Fig. 13.4. The build mesh is represented in orange with two settings meshes represented in purple. In this example, the settings meshes increase the infill density of the areas they cover. So, when this object is constructed, the area around the holes will have more material, which is ideal if the holes were expected to be tapped for threading. Similarly, for the cross-shaped object in Fig. 13.3, a square settings mesh was placed around the center of the object. That settings mesh expanded the melt pool by increasing the laser power, and subsequently resulted in a successful build.

Advanced site-specific control

A logical extension of settings meshes is the manipulation of toolpaths such that excess material is deposited in specific areas of an object to create an embossing on one surface. However, such an embossing process needs more specific control than that provided by meshes that cover entire areas of the build space; it requires settings meshes that are localized to areas within a build mesh. This degree of specificity can be implemented by developing a parent-child relationship for build and settings meshes, respectively. This parent-child relationship allows child settings meshes to manipulate the settings of a specific build mesh (its parent), rather than just an entire region of the printer's build space. Additionally, there is no reason that a child setting mesh cannot be contained within its parent build mesh. Orienting such a child mesh upon the surface of a parent mesh allows the distinct construction of the child mesh through manipulations of settings alone, without explicit pathing. An example of this is shown in Fig. 13.5. In this figure, the orange object represents a simple wall and is the parent object, while the purple leaf is the child settings mesh and is embedded into the surface of the parent mesh.

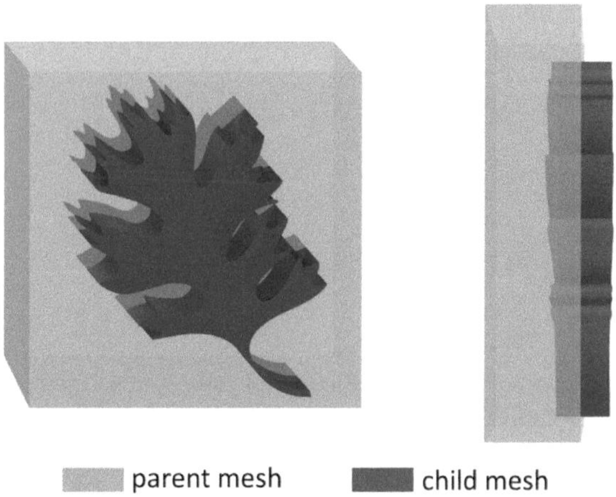

FIG. 13.5 Parent-child setup. The parent is the orange wall that will have pathing computed as normal. The purple child object is a leaf that will be intersected with its parents' pathing to modify construction parameters.

Like the site-control in the previous section, the child mesh modifies the pathing of the parent mesh. If the child mesh alters the settings to expand the melt pool by 50%, areas where the pathing of the parent mesh intersects with the child mesh will be printed with 50% more material. Conversely, pathing that does not intersect the child is left under normal build settings, which relatively deposits less material. Switching between these two melt pool states produces an uneven surface that is wider in areas that intersect with the child mesh. The result is an embossed shape on the surface of the printed object, as shown in Fig. 13.6. Note that the resulting printed object expands beyond the geometry of the original parent. However, because we are manipulating in the *XY* plane, this difference from the expected object geometry does not interact with height control.

In this example, pathing was only computed for the parent wall object and the leaf shape was printed just by manipulating the settings of sections of the parent's pathing. This is obviously a very simple example. The parent wall was a simple geometry and, therefore, the pathing was simple as well. However, this is just a proof of concept. There is nothing that prevents this approach from being applied to non-flat surfaces, examples of which are shown in Fig. 13.7. Additionally, it would be possible to deboss a shape onto an object's surface by using the same method to shrink the melt pool and reduce material deposition in specific areas. There are in fact many creative applications for child meshes. This section has only served to illustrate how advanced site-specific control can be achieved by a slicing solution.

FIG. 13.6 Result of using settings meshes for site-specific control. This object shows an embossed leaf on the surface of a wall before (left) and after (right) polishing. *Credit: B.T. Gibson, Y.K. Bandari, B.S. Richardson, W.C. Henry, E.J. Vetland, T.W. Sundermann, L.J. Love, Melt pool size control through multiple closed-loop modalities in laser-wire directed energy deposition of Ti-6Al-4V, Addit. Manuf. 32 (2020) 100993.*

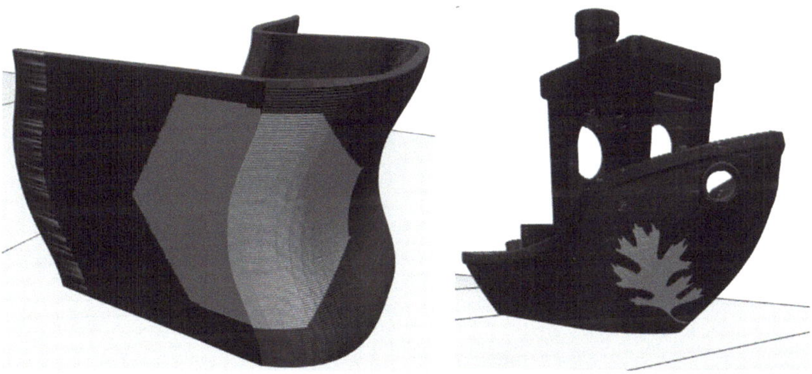

FIG. 13.7 Examples of the embossing process using parent-child meshes on nonflat surfaces. *Credit: M. Borish, B.T. Gibson, C. Adkins, P. Mhatre, Automated process planning for embossing and functionally grading materials via site-specific control in large-format metal-based additive manufacturing, Dent. Mater. 15(12) (2022) 4152.*

Conclusion

This chapter has introduced some basic considerations when pathing for DED systems. In general, the melt pool dynamics of DED systems make construction a challenge and are a research topic of their own. Unlike polymer systems, DED systems need control over layer height to produce successful, high-quality builds because the deposited metal is a liquid until it cools. However, manipulation of this liquid melt pool provides unique opportunities for control over the printing process. For example, voids left by paths can be filled through site-specific control of material deposition, and surfaces of objects can be embossed or debossed through advanced site-specific control. Basic slicer support for such capabilities was described in this chapter. This chapter also serves as the foundation for advanced research topics discussed in *Chapter 19: Closing the loop*. As mentioned in the introduction, DED systems also have additional degrees of freedom due to the use of robotic arms instead of gantries. Pathing nuances related to those additional degrees of freedom will be touched on in *Chapter 16: Pathing considerations for 5-axis systems*.

References

[1] M. Borish, B.K. Post, A. Roschli, P.C. Chesser, L.J. Love, K.T. Gaul, Defect identification and mitigation via visual inspection in large-scale additive manufacturing, JOM 71 (3) (2019) 893–899.
[2] B.T. Gibson, B.K. Post, M.C. Borish, A.C. Roschli, Y.K. Bandari, B.S. Richardson, A. Thornton, W.C. Henry, L.J. Love, Multi-Modal Sensing and Control in Laser-Wire Based Directed Energy Deposition, DOE-NNSA Additive Manufacturing Workshop, Knoxville, TN, 2019.
[3] B.T. Gibson, Y.K. Bandari, B.S. Richardson, W.C. Henry, E.J. Vetland, T.W. Sundermann, L.J. Love, Melt pool size control through multiple closed-loop modalities in laser-wire directed energy deposition of Ti-6Al-4V, Addit. Manuf. 32 (2020) 100993.

CHAPTER

14

Planar slicing for nonextrusion AM processes

Eric MacDonald[a], *Michael Borish*[b], *Abby Barnes*[b], *Alex Roschli*[b], *Peter Wang*[b], and *Thomas Feldhausen*[b]

[a]University of Texas at El Paso, El Paso, TX, United States, [b]Manufacturing Demonstration Facility, Oak Ridge National Laboratory, Knoxville, TN, United States

Abstract

Extrusion-based AM processes, including material extrusion and directed energy deposition, construct objects by continuously depositing a relatively small amount of molten feedstock to a specific location. Nonextrusion AM processes, on the other hand, tend to form each layer by linearly or areally projecting either energy onto a vat of photocurable liquid or a binding agent into a bed of loose powder. Such systems vary significantly from extrusion-based AM processes but do share similarities. The general approach to path planning is the same, but nonextrusion processes typically favor the use of bitmap representations, instead of polygons, for pathing computation and do not use a g-code file to command the printer. In this chapter, a high-level discussion of these differences will be covered.

Introduction

So far, this book has focused on slicing and path planning for extrusion-based AM processes, all of which involve the dynamic deposition of the desired material. However, not all AM processes use this approach to layer-based construction. Some AM processes selectively harden or bind specific locations of a bed or vat of the desired material, instead of depositing the material on a surface, and are therefore collectively called "non-extrusion" processes.

https://doi.org/10.1016/B978-0-443-15286-3.00008-4

The collection of non-extrusion AM processes includes vat photopolymerization, binder jetting, and powder bed fusion, which were all briefly covered in *Chapter 1: Introduction to additive manufacturing*. Importantly, the motion and controls of non-extrusion systems are different from those of extrusion systems, as was mentioned in *Chapter 3: Motion of platforms*. Non-extrusion AM systems tend to have a single degree of freedom, meaning motion is limited to one axis, instead of having motion in the X, Y, and Z directions. Usually, the motion of non-extrusion systems is limited to raising and lowering a bed of material along the Z axis. Targeting along the X and Y axes is achieved either by angling mirrors to direct a beam of energy, or by selectively toggling multiple energy or binder sources on or off.

Due to these differences in motion platforms, non-extrusion processes require a slightly different approach to path planning. This chapter will highlight aspects of slicing and path planning that are unique to non-extrusion AM processes. It is important to note that there are many different styles of non-extrusion systems, and each has a slightly different method of path computation. What is described in this chapter is a generalized approach to path planning for non-extrusion systems, with a handful of details related to specific styles of non-extrusion systems, such as metal powder bed and binder/ink jet systems.

Compared to extrusion-based systems, the specifics of path computation for non-extrusion systems are less publicly known; they are often machine-specific and obscured by the manufacturer. At the time this book was written, multiple well-known, open-source slicers existed for extrusion systems. The same was not true for non-extrusion systems, whose software tends to be more commercialized, and as a result, proprietary. Therefore, there is much less available information on slicing for non-extrusion, which is why this book only has this single, brief chapter on the topic.

The basics of slicing for non-extrusion

The first step of any slicing process is cross-sectioning, and slicing for non-extrusion is no exception. In general, cross-sectioning for non-extrusion systems is the same as those for extrusion systems. A full discussion of cross-sectioning can be found in *Chapter 5: Cross-sectioning*. At a high level, the process takes the "triangle soup" that composes an STL and attempts to intersect each triangle with a plane at a given height. Based on the intersection, segments are identified and combined to form the boundary contours of a polygon. In certain respects, cross-sectioning for non-extrusion systems is easier than extrusion systems. This is because the build volume for non-extrusion systems tends to be much smaller and the resolution of such systems tends to be much higher. This means, for example, that there is no need to smooth cross-sections, as was detailed in *Chapter 5: Cross-sectioning*. There is no large machinery to move, so large, fluid motions are not a concern. However, that does not mean that smoothing cannot be applied, just that it is usually unnecessary.

The major difference between slicing for non-extrusion vs. extrusion occurs after cross-sectioning, during path generation. One major characteristic that sets slicing for these systems apart is the use of bitmaps in the pathing process. A bitmap is a spatial representation of "bits" of information and is typically used to create images on computer screens. A bitmap mask, which can also be called a "bitmask," is a specific form of bitmap that contains opacity information. A bitmask can be used to control which parts of an image are displayed (i.e., have an opacity above zero) and which parts are not (i.e., zero opacity). Bitmasks are generally the type of bitmap used in non-extrusion path planning. They are inherently greyscale, with black being fully transparent and white being fully opaque. However, it is common for many systems to ignore this opacity and treat the mask as binary—being either fully black or white at any location.

Although the details of the pathing generation process for each non-extrusion machine are slightly different and largely unknown, one thing is true about all of them: they use bitmaps instead of g-code. Non-extrusion machines don't use g-code because, as mentioned before, their motion and controls are often very different from extrusion machines. However, exactly when and how the cross-sectioned information gets converted to a bitmap is different for each non-extrusion process. As such, the type of data that each machine uses to construct an object varies widely, leading to a large variety of file types that are produced by the slicing process.

Data representation

Despite the variety of file and data types used by non-extrusion systems, there seem to be commonalities that divide non-extrusion slicing processes into three categories: those that output a vector-based file, those that output a bitmap, and those that don't involve a slicer at all. However, although those categories seem distinct, they are actually variations on a theme. For example, systems that don't require a slicer are not systems that don't slice 3D models. Rather, they are systems that read an STL file and include the slicing process as part of the file processing. These machines ultimately still generate bitmaps from the STL files. Many modern slicers for binder jet follow such an approach by utilizing the machine's graphics processing unit (GPU) and associated graphics libraries to rapidly produce layer bitmasks. Similarly, machines that read vector-based files produce bitmaps from those vectors as part of their file processing.

To illustrate the non-extrusion slicing process, this chapter will summarize slicing that produces a vector-based file. Like bitmaps, vector files can be used to render images, but instead of a map of information, a vector file describes a series of points and their connections using a sequence of commands or mathematical statements that place lines, and fill between them, in a two-dimensional space. For non-extrusion slicing, vector-based output is the most indirect route to a bitmap. Therefore, it provides an example of how an STL file can be sliced and converted into bitmaps that are then used to construct objects. Remember, though, that this process is not universal. Many non-extrusion machines have their own unique slicing process.

The CLI file: An example of slicing for nonextrusion

The CLI file is a vector-based file that was developed with funding from the Basic Research in Industrial Technologies for Europe/European Research on Advanced Materials program (BRITE EuRAM 1994) with the intent to produce a simple, efficient, and unambiguous format for planar geometric data. It is based on a "2.5-dimenion" ("2.5D") layer representation, in which layers are rendered in 2D, but are organized such that the third dimension (usually Z) can be inferred based on the location of each image relative to the other images on the bitmask.

Many current commercial non-extrusion systems utilize a CLI file or CLI-like file format as an intermediate file prior to the generation of system-specific control data structures. The required elements of the CLI language are application-independent, but the USERDATA command allows the generation of application-specific information. This flexibility allows the format to be used for a diversity of applications without loss of crucial geometry data. In the context of AM, a CLI file uses proprietary internal data structures to command systems [1–3], making it suitable for use by commercial providers of AM equipment. Recently, CLI was leveraged to inspect additively manufactured parts during fabrication, layer by layer, for in situ qualification [4,5].

A CLI file encodes the contours of a polygon that has been computed from cross-sectioning a 3D geometry. Fig. 14.1 shows a 3D model and its corresponding layers of contours that would be encoded in a CLI file. The 3D model shown in Fig. 14.1A is an industrially relevant turbine that includes sufficiently complex geometric features to demonstrate the slicing process for 9 layers, as shown in Fig. 14.1B [6]. The contours in Fig. 14.1B show the edges of the 3D model at various cross-sections, starting from the bottom of the geometry and working upwards. Each layer is arranged in proceeding order from left to right, top to bottom.

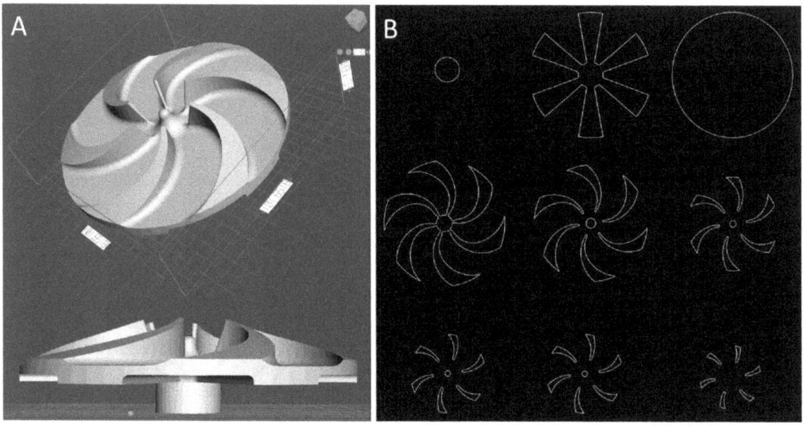

FIG. 14.1 (A) Example geometry with dimensions (top) and a side view (bottom); (B) layer bitmasks for the example geometry, in which the upper left corner represents the bottom of the geometry and the lower right corner represents the top of the geometry. The bitmasks proceed left to right, top to bottom. *Credit: [1]*

A CLI file can be used to define boundaries and detect both external and internal contours because they contain sets of vertex points (usually XY coordinates) that can be connected contiguously by straight line segments. These straight line segments form a closed polyline, which can also be called a polygon and is essentially the same as the polygons discussed in *Chapter 5: Cross-sectioning*. Just as with the previously discussed polygons, the clockwise or counterclockwise order of the points in the polygon determines if the line defines an external boundary (counterclockwise) or the edge of an internal cavity (clockwise).

Variables in the CLI file define the units used (millimeters by default) as well as the bounding box of the entire structure within the plane. Designators indicate the beginning and ending of geometric features and the layers within each feature. Each layer has a defined thickness, which can be unique to that layer because layer thicknesses are not necessarily uniform. Beneath layer designators are lines of code to define the polylines within that layer. The code for polylines is a list of numbers that starts with an ID to define which object the polyline belongs to, followed by a number that indicates the type of polyline (0 for internal contour, 1 for external contour, or 2 for open line), then the number of vertices in the polyline, and finally the X and Y values for each of the vertices. Fig. 14.2 shows an example as provided by the original Common Layer Interface documentation [3]. Just like the STL files discussed in *Chapter 4: Geometry Data Storage*, CLI files can be stored in either the human-readable ASCII format or in the less readable and but significantly smaller-sized binary format.

Once an AM system ingests a CLI file, computer graphics libraries within an AM system must generate a bitmask whose opacity values tell

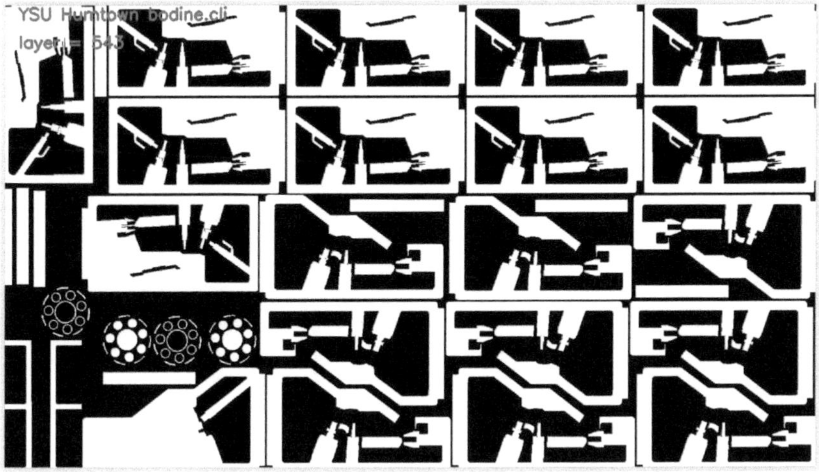

FIG. 14.2 Example slice of a 3D printed sand mold fabricated with sand binder jetting.

the system where to solidify the material within the bed or vat, and where to leave the material unsolidified. Opacity information can be used to convey on and off commands to a grid of binder or energy sources; areas that are transparent (black) indicate the source should be off, while areas that are opaque (white) indicate the source should be turned on. When used this way, the bitmasks produced from the CLI file are interpolated and discretized to form a single image that is then used to appropriately toggle the binder/energy sources as layers are constructed. An example of a bitmask produced by reading a CLI file is shown in Fig. 14.2.

One shortcoming of the CLI format is that it has no grayscale capabilities, which is apparent in Fig. 14.2. Most non-extrusion processes today use simple binary bitmasks but, conceivably, shades of gray in a bitmask could be used to deliver more or less binder to allow easier cleaning of printed objects or to facilitate the removal of support structures. Grayscale could also be used to improve surface finish, increase dimensional accuracy, or vary other aspects of the printed part, such as color. In fact, grayscale imaging has been recently used to improve the surface finish and dimensional accuracies of parts printed via vat photopolymerization, even though grayscale is not currently captured in the CLI format [7–9].

Using the data

The above example illustrates how a machine would create a bitmap by processing a CLI file. For machines that directly read bitmaps, the slicing process would skip the production of the CLI file and just produce a bitmap like the one in Fig. Fig. 14.2. For machines that read STL files, the entire process (or a similar process) is included within the machine's proprietary processing.

Regardless, once the conversion to a bitmap is complete, pathing can be approached in a manner that is similar to extrusion systems. The shape represented in the bitmap can be fit with closed-loop contours, which are generated based on various user settings and were described in *Chapter 6: Closed-loop toolpath generation*. Non-extrusion systems also use the space-filling toolpaths discussed in *Chapter 7: Space-filling toolpath generation*. Again, the difference is in how the machine is commanded to move along the generated paths. For some systems, instead of moving a deposition head around the XY plane, the angle of mirrors might be changed, or various members of a set of binder sources might be toggled on or off.

In general, prints done by non-extrusion systems involve a single closed contour, or perimeter, that bounds the object while the remaining area is filled with infill, also called hatching. At a high level, hatching is typically a series of parallel lines at some angle that fill the interior space of the object, similar to the rectilinear pattern discussed in *Chapter 7: Space-filling toolpath generation*. Additionally, just as with extrusion AM, there is a wide variety of infill patterns used in non-extrusion AM.

Pathing for non-extrusion systems typically does not involve open-loop toolpaths because voids are not a concern. The small build volumes and very high resolutions of these systems generally make extremely small voids. As a result, pathing solutions for such systems typically do not include open-loop paths in their computations. This, along with the reduced need for smoothing that was mentioned earlier, makes pathing for non-extrusion AM relatively simpler than pathing for extrusion-based processes.

Non-extrusion systems also have unique characteristics when dealing with various pathing optimizations, but in this case, these characteristics alter the source of complexity rather than just simplifying the process. Whether this change in complexity is positive or negative depends on the objects intended for construction. As with all AM, certain processes and pathing styles lend themselves to the construction of certain objects.

One example of the altered complexity of non-extrusion systems is the computation of travels. Travels are motions between contours or build paths in the construction of an object and were discussed in detail in Chapter 12: Pathing considerations for extrusion: Pellet, filament, concrete, and thermoset. Travels allow repositioning of the machine, to restart construction in a new location. In general, non-extrusion systems do not use travels because repositioning is either implicitly handled by the system controller or it is completely inapplicable—another reflection of their motion platforms. However, with travels disregarded, new considerations arise for how paths can be optimized, and these new considerations bring additional complexity.

For example, powder bed systems tend to have travels that are implicitly controlled, because each polyline is a separate output command. The end effector automatically jumps from polyline to polyline without explicit commands to do so, and for all practical purposes, these jumps are considered instantaneous because the optics can be repositioned very quickly. However, these motions can be manipulated by shuffling polylines into different orders, adding a unique level of complexity. Sometimes "point melting" approaches are used, in which electron beam systems jumps around to separate points to construct a layer, rather than following a path of segments. This is usually used to control the properties of the melted material.

For inkjet printers, travels are completely irrelevant. In these processes, the material can be thought of as extremely small droplets. The systems deposit these droplets in specific locations as the scan head moves across the entire build surface in a single motion. Imagine filling in pixels across the entire length of an image. There are no travels because nothing ever needs to be repositioned. However, many "pixels" can be filled simultaneously and not necessarily with the same material. Although this seems simple, optimizing the processing and execution of this simultaneous activity can become very complex.

Conclusion

In this chapter, some high-level differences between path planning for extrusion and nonextrusion-based processes have been touched on. The major differences are how geometric data is represented and how paths need to be optimized. Instead of maintaining polygons as the representation of geometry, nonextrusion AM systems convert geometric information into bitmasks. The pathing of these bitmasks does not usually involve computations for open-loop paths or travels but may require computations for the simultaneous deposition of multiple materials and/or very quick (almost instantaneous) and relatively large jumps in the location of deposition. However, despite these differences, the fundamental approach to produce pathing remains the same.

The output of the slicing process for nonextrusion systems varies, but is often a vector-based file such as a CLI file. The CLI and similar formats are currently used to generate bitmasks whose binary (black or white) values command deposition in specific locations, at a defined rate. Opportunities exist to optimize nonextrusion processes by transitioning to a nonbinary, grayscale bitmask. Such a nonbinary data representation could provide new capabilities to these AM processes, including varying the strength, density, and color of the material within a single part.

References

[1] Mueller T. How will 3D metal printing impact investment casting? In: How will 3D Metal Printing Impact Investment Casting? [Internet]. Available: http://www.google.com/url?sa=t&rct=j&q=&esrc=s&source=web&cd=4&ved=0ahUKEwi21YnqxsHbAhUkwlkKHY7SDIoQFghBMAM&url=http%3A%2F%2Fweb.investmentcasting.org%2FExternal%2FWCPages%2FWCWebContent%2FWebContentPage.aspx%3FContentID%3D422&usg=AOvVaw0Yt0J8UeLxft6UIRtsdFyl.

[2] F. Mandorli, U. Cugini, V. le delle Scienze, Identification of process relevant form structures for stereolithography within solid models, Proc IFIP WG5 (1996) 313–324.

[3] C.C. Kai, G.G.K. Jacob, T. Mei, Interface between CAD and rapid prototyping systems. Part 1: a study of existing interfaces, Int. J. Adv. Manuf. Technol. 13 (1997) 566–570.

[4] A. Welisch, Common Layer Interface Documentation. In: Common Layer Interface [Internet], 2000. [cited 29 Dec 2022]. Available: https://www.hmilch.net/downloads/cli_format.html.

[5] S. Clijsters, T. Craeghs, S. Buls, K. Kempen, J.-P. Kruth, In situ quality control of the selective laser melting process using a high-speed, real-time melt pool monitoring system, Int. J. Adv. Manuf. Technol. 75 (2014) 1089–1101.

[6] T. Craeghs, S. Clijsters, E. Yasa, J.-P. Kruth, Online quality control of selective laser melting, in: Proceedings of the Solid Freeform Fabrication Symposium, Austin, TX. Citeseer, 2011, pp. 212–226.

[7] A. Andreu, P.-C. Su, J.-H. Kim, C.S. Ng, S. Kim, I. Kim, et al., 4D printing materials for vat photopolymerization, Addit. Manuf. 44 (2021) 102024.

[8] E. Guven, Y. Karpat, M. Cakmakci, Improving the dimensional accuracy of micro parts 3D printed with projection-based continuous vat photopolymerization using a model-based grayscale optimization method, Addit. Manuf. 57 (2022) 102954.

[9] G.I. Peterson, J.J. Schwartz, D. Zhang, B.M. Weiss, M.A. Ganter, D.W. Storti, et al., Production of materials with spatially-controlled cross-link density via vat photopolymerization, ACS Appl. Mater. Interfaces. 8 (2016) 29037–29043.

CHAPTER

15

Off-axis and nonplanar slicing

Michael Borish[a], *Alex Roschli*[a], *Abby Barnes*[a], *Eric MacDonald*[a,b], *Peter Wang*[a], *and Thomas Feldhausen*[a]

[a]Manufacturing Demonstration Facility, Oak Ridge National Laboratory, Knoxville, TN, United States, [b]University of Texas at El Paso, El Paso, TX, United States

Abstract

Traditionally, slicing processes have assumed a horizontal plane that moves up the *Z* axis to produce layers. However, through relatively straightforward modifications, this process can be extended to allow planar slicing at any angle and along any axis in a process referred to as off-axis slicing. Additionally, more advanced pathing solutions are moving away from planar approaches entirely by printing directly on nonflat surfaces. Both of these alternative approaches to slicing, as well as some common issues associated with them, are discussed in this chapter to provide a foundation for more advanced pathing solutions.

Introduction

The previous chapters have detailed the processes used to generate the necessary pathing to print horizontal layers that incrementally rise along the *Z* axis. In this chapter, two additional pathing approaches will be described: off-axis and nonplanar. Off-axis printing is the name for 3D-printing layers that are angled to intersect the horizontal plane, instead of being parallel to it, and requires the support of off-axis slicing. Nonplanar printing, on the other hand, is the name for 3D-printing with a machine that has the ability to print at varying angles and thus entirely disregard the horizontal plane. Both processes can leverage the work of the previously described slicing process, so that there is no need to start from scratch. Planar construction via off-axis slicing will be discussed first, followed by the conformal mapping approach to nonplanar printing.

https://doi.org/10.1016/B978-0-443-15286-3.00017-5

Off-axis printing

Before discussing how to generate off-axis pathing, it's important to discuss the motivation for such an approach. As was mentioned in *Chapter 9: Support, raft, brim, and skirt pathing*, support structures can be used to construct an object that has a significantly overhanging feature. However, the use of supports increases the total build time and the material used for the print. Moreover, as objects scale in size, the use of supports becomes less viable. For a desktop system, supports can be broken off by hand and the surface cleaned with a piece of sandpaper, but for a large format system, supports must be removed with the use of power tools or machining equipment. This is tedious, time-consuming, and potentially dangerous, making it undesirable. If the object is constructed at an angle that is not parallel to the horizontal axis, however, then the overhanging feature becomes achievable without the use of support structures.

Another key motivation for off-axis printing is the size of 3D printers. Again, large format systems can produce very large parts, but there is always an increasing need to print even larger and more complex objects. It is unreasonable to assume that existing machines will simply continue to scale up, as hardware limitations prevent this. In response, next-generation systems with significantly different hardware are rising in popularity. An industrial example is SkyBAAM [1]. This is a concrete 3D printer that is cable-driven. As a result, there is no statically defined build volume. Another relevant example would be the rise of belt-driven 3D printers, which were discussed in *Chapter 3: Motion of platforms*. These are usually smaller in scale but seek to allow infinite construction along one axis of the system. This construction is accomplished by use of a conveyor belt instead of a standard build table, as shown in Fig. 15.1. By using a belt, the object is continuously moved along one of the axes of the system and can even be moved outside the volume of the 3D printer. This axis is typically the X axis, and so, by printing at an angle, objects that are much longer than the system can be produced.

To allow off-axis printing, alterations must be made to the cross-sectioning and optimization algorithms of *Chapter 5: Cross-sectioning* and *Chapter 11: Travels, optimizations and ordering*. However, with intelligent manipulation of planar information, the entire pathing process described in the intervening chapters can remain unchanged. Recall from *Chapter 5: Cross-sectioning* that cross-sectioning for a horizontal plane requires comparing a layer's height (its Z coordinate) to the vertices of the triangles of the STL, to see if and where the layer intersects a triangle. This is not sufficient for off-axis cross-sectioning. Instead of comparing just the Z-axis component, now a representation of the cross-sectioning plane must be created, and the vertices of the triangles must be evaluated against this plane.

There are multiple ways to create a representation of a plane in computer software. One of the simplest is the point-normal form, which uses

FIG. 15.1 A small format belt-driven system. In this example, the belt is oriented with the X-axis. This allows the machine to print very long parts when slicing and printing at an angle.

a single fixed point on the plane to create a vector that is normal to the plane and can be used represent the plane. The point used in the representation can be any point on the hypothetical plane and the normal vector can extend in two different directions, but for the purposes of off-axis slicing, the chosen point is the centroid of the object and the normal will be directed towards the top of the object. Unlike the cross-sectioning in *Chapter 5: Cross-sectioning*, using this representation allows the use of unique cross-sections on a per-part basis. That is, if multiple objects are being constructed, each can be cross-sectioned and built at a different angle. This capability is one of the foundations of more advanced pathing solutions.

With the plane well-defined, the cross-sectioning begins by starting at the minima of the object, relative to the slicing plane. In most cases, this means that the first cross-section is not the bottom of the object but is instead a corner, as the plane is shifted to get to the first valid cross-section of the object. To produce a cross-section, each pair of vertices on a face is used to construct a line segment, and this line segment is tested against the plane to see if the plane intersects it. This test, shown in Fig. 15.2, is analogous to the cases in *Chapter 5: Cross-sectioning*: a point will only be inserted when one vertex is above and one vertex is below the plane.

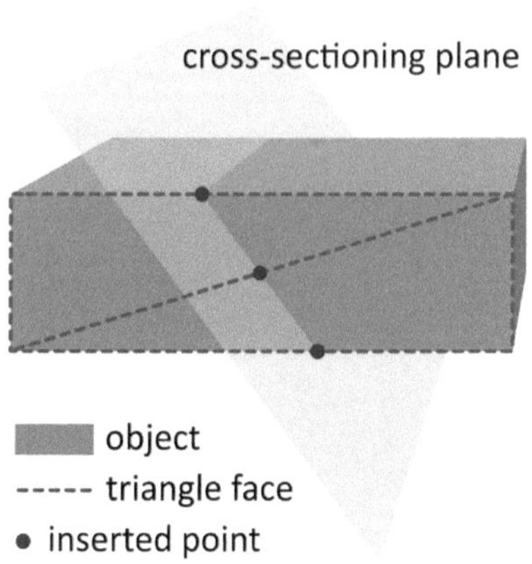

FIG. 15.2 Example point on the line segment generated by the intersection of the angled plane with the object. This intersection test will include this line segment and create an angled, square cross-section.

One important note is that, unlike when using a horizontal cross-section, the *Z* coordinates of any generated points must be computed and stored. Remember that, when using a typical horizontal cross-section, the *Z* components of all the resulting polygons and pathing can be dropped to speed up computation, which was also mentioned in *Chapter 5: Cross-sectioning*. The *Z* coordinate is not necessary because all pathing for a particular cross-section will have the same value, based on the layer height of the cross-section. In contrast, when using an off-axis or nonplanar cross-section, not all points in a layer with share the same *Z* coordinate, so each point's *Z* coordinate needs to be maintained.

Once all intersection points have been found for the current cross-section, all points are rotated to an equivalent horizontal plane. This can be accomplished using either a quaternion or a rotation matrix [2,3]. These were discussed in depth in *Chapter 2: Robot kinematics*, but at a high level, they are very similar and either can be used to represent a rotation in 3D space. They perform identically in many situations, and both are usually well-supported in any modern programming language. In any case, whichever one is selected will be used to represent the rotation of the points generated from the cross-section to equivalent points on a horizontal plane. This rotation is illustrated in Fig. 15.3. Importantly, this rotation should be calculated from the centroid of the object. Otherwise, the points would need to be translated to an origin for correct computation of the rotation. Conveniently, because the cross-sectioning plane is described by the centroid in the point-normal representation, that information is at hand for the calculation.

These steps are, in fact, just a transform back to the original horizontal slicing problem. This is because two planes can be rotated to align with

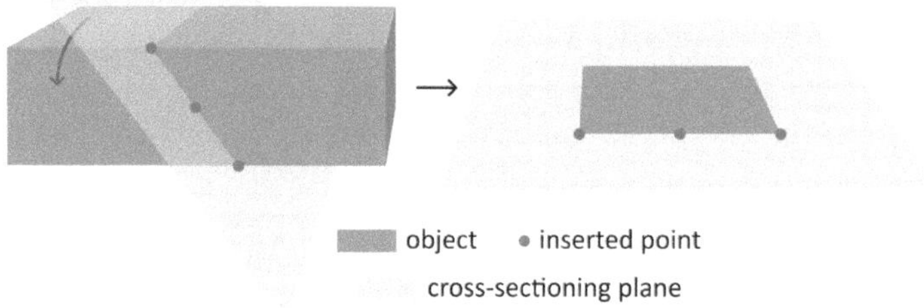

FIG. 15.3 Rotation of the off-axis cross-sectioning plane. Once the segments that compose the angled cross-section are determined, they are rotated to a horizontal plane. This causes all points to be equivalent to a traditional cross-section, allowing use of the previously discussed algorithms.

one another without loss of information or the introduction of deformation. Each point will need to carry along this rotational information during the computation process, but the process can proceed as normal. The off-axis points have now been transformed into a flat cross-section, and as a result, the entire slicing process described in *Chapter 6: Closed-loop toolpath generation*, *Chapter 7: Space-filling toolpath generation*, *Chapter 8: Open-loop toolpath generation*, *Chapter 9: Support, raft, brim, and skirt pathing*, and *Chapter 10: Path modifiers* can be leveraged to generate appropriate pathing. Additionally, although the *Z* component of each point must be calculated and stored, it can be ignored while generating toolpaths to speed computation.

With one cross-section complete, the plane is then shifted based on the layer height, just as the traditional cross-sectioning plane is raised based on the layer height. The typical behavior is to shift the plane along its normal and iteratively take cross-sections until the plane passes the top of the object. This is another reason the point-normal form for the cross-section plane is used. However, there is no reason that the shift in the plane needs to be along the normal. The shift could just as easily be along any other axis. In this way, the angle and motion of the plane can be decoupled and allow cross-sectioning at any angle along any axis.

The flexibility of this approach does mean that it is more complex and, as such, comes with additional concerns that must be addressed. For example, the most significant issue will be combinations of cross-sectioning planes and angles that are invalid. This occurs when the angle of the cross-section plane is parallel to the vector representing the motion of the plane. Consider the case in which a cross-sectioning plane is horizontal, if such a plane were used to cross-section towards the *X*-axis, the cross-sectioning would never stop. A plane is infinitely large and so will always produce valid geometry. Moving the angled plane along its normal prevents this issue, which is why that is the usual approach.

Once cross-sectioning is complete, the pathing process continues to optimizations and travel insertions as usual. These are computed as normal, but after they are completed, the resulting pathing is rotated back into place from its horizontal representation. This is accomplished through the inverse of the rotation that the points have been carrying during the computation process. After the transform is complete, the pathing can have its g-code generated as usual. Naturally, to generate the appropriate g-code, the Z coordinates of each point must be incorporated into motion commands. However, this is a small extension to already existing g-code generation methods, which are covered in *Chapter 18: The G-code file.*

Printing via automatic off-axis computation

One natural extension to off-axis computation is the ability to automatically determine the cross-sectioning plane. This approach has been motivated by systems that have four or more degrees of freedom. For a system with four degrees of freedom, the first three degrees of freedom are provided by a typical gantry or robotic arm for X, Y, and Z motion, while the fourth degree of freedom is the result of a tilt table. As the name implies, a tilt table can tilt along one or more axes. By having the ability to tilt, the angle of the object being built can be changed, relative to the extruder, during construction. This enables the construction of complex geometries and significant overhangs without the need for support pathing and material. These types of systems are intermediary to the more common 5-axis systems that will be discussed in more detail in *Chapter 16: Pathing considerations for 5-axis systems*. Both 4- and 5-axis systems can benefit from automatic computation of the angle of the cross-sectioning plane.

To accomplish this automatic computation, the first step is to skeletonize the mesh. Skeletonization of a mesh is similar to the skeletonization of a polygon region that was discussed in *Chapter 8: Open-loop toolpath generation*. The mesh is decomposed into a representative skeleton that maintains the topology of the geometry. Additionally, just like skeletonization of a polygon, there are multiple approaches to generating the skeleton of a mesh. One approach is the use of mean curvature flow (MCF). MCF is typically used to smooth meshes by reducing the area of the mesh's surface but has several characteristics that are also useful for skeleton generation. Chief among those characteristics is the representation of minimal surfaces, which can be used to shrink a mesh into a topologically representative skeleton that maintains the curvature of the mesh [4]. This process allows skeleton segmentation, which is useful for complex geometries such as the one in Fig. 15.4. The object in the figure has clear regions, indicated by color, that the skeleton can be segmented into. This segmentation allows one "tube" to be built in its entirety before moving to the next

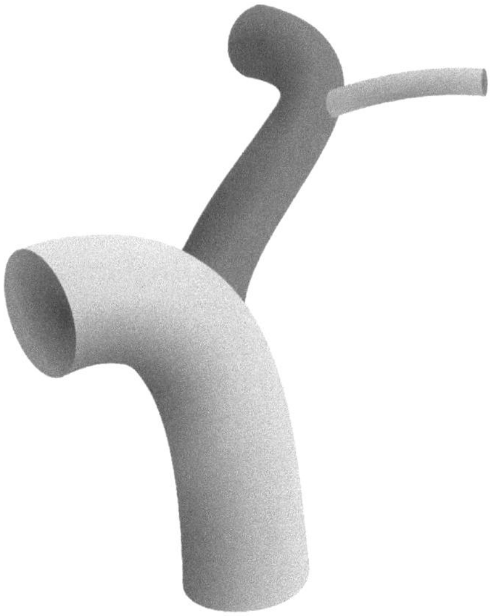

FIG. 15.4 Example of automatic off-axis computation. In this example, each area of the mesh highlighted in a different color is segmented. Each area has its skeleton computed with the slicing plane aligning to that skeleton. Each area is built in order, starting at the bottom-most segment and going to the next connected piece.

one, which avoids collisions and allows the planar tilt angle to be calculated independently for each subsection.

Once the points of the skeleton are generated, the next step of the MCF approach is to fit a Bezier curve to them. Then, this curve is evaluated at intervals of the desired layer height and, at each interval step, a normal is calculated to determine the appropriate angle of the slicing plane. The behavior of the slicing plane along this normal is like that described for off-axis slicing: the edges of the faces of the STL are evaluated for intersections with the plane, points are inserted, the points are transformed to equivalent points on a horizontal plane, pathing is generated, and the transform is reversed. With this process in place, a tilt table can be leveraged to construct objects with various overhanging features without the need for supports. As *Chapter 16: Pathing considerations for 5-axis systems* will show, more advanced solutions are possible with the transition to a 5-axis system.

Nonplanar printing via conformal mapping

In addition to off-axis pathing, other nonplanar pathing solutions also exist. Some nonplanar pathing solutions allow the deposition of material onto curved and/or complex surfaces, instead of on a flat, planar surface such as a build plate. Many of these approaches are based on texture mapping algorithms that are regularly used in 3D graphics. Texture mapping refers to processes that map a 2D texture to a 3D object. This is typically accomplished by "wrapping" the 3D object, much in the same way that a gift box may be covered with wrapping paper.

One common approach used is conformal mapping [5]. A conformal map parameterizes a 3D object to a 2D planar surface—or, in other words, maps the 3D information onto a 2D plane. Conformal mapping attempts to minimize angle distortions at the cost of area or length distortions. This means that a set of parallel lines would remain parallel when mapped but may be longer or shorter than in the original 3D object. This is convenient for 3D-printing on a surface; the build paths may need to be longer or shorter to cover a surface but would remain aligned as if building on a plane.

It should be noted there are multiple ways to map a nonplanar to planar surface, but all mappings will have distortion. A very familiar analogue is converting a globe to a map—a 3D globe is functionally mapped to a 2D plane. Each style of map can appear significantly different due to the different distortions inherent to the chosen method of mapping.

The first step in the conformal mapping process is to determine the surface that will be built upon. This surface can be a portion of an object and will be transformed to a planar representation. For any reasonably complex geometry, not all surfaces will be accessible by the system, and so, only specific surfaces can be used for construction. One straightforward approach for identifying such a surface is by simple normal evaluation. If a triangular face's normal is upward/outward facing, then it can be mapped to a planar surface. Once this surface of interest has been determined, a transformation function must be computed. This transformation will map the 3D surface pointwise to a 2D plane, allowing the surface to be "squished" and flattened. This is easily illustrated with a simple dome. When a dome is conformally mapped, the result is a flat circle.

However, this planar representation exists in a different coordinate system, UV, rather than cartesian XYZ space that the object and the build volume exist in. U and V typically correspond to X and Y, respectively, in 3D (XYZ) space. UV space, unlike 3D space, is unitless, bounded, and relative. U and V coordinates indicate their position relative to the boundaries of the space and, as such, have a minimum of 0 (at one edge of the space) and a maximum of 1 (at the opposite edge of the space). This presents a challenge because UV space does not account for any of the intricacies necessary for 3D printing an object, such as spatial location or units of length.

To utilize the pathing approaches of previous chapters, a boundary polygon for the surface must be extracted. Using the vertices that represent the "edges" of the surface of interest, a 2D bounding polygon can be extracted from the conformal mapping. This polygon can then be used with the standard pathing approaches outlined in previous chapters. An example of this conformal mapping process is shown in Fig. 15.5.

Once the necessary pathing has been produced, like the off-axis approach, the planar pathing must be transformed again via an inverse of the original transformation process (in this case, conformal mapping) to arrive at a 3D pathing solution. However, there is one more significant issue related to the resolution of the pathing as it is passed through the conformal mapping algorithm. Consider again the dome from Fig. 15.5. When computing the 2D line infill pathing, the generated paths are single segments spanning the entire circle. If these segments were remapped in 3D space, the result would remain flat because there is no way to assign multiple Z-axis coordinates to a single path.

To deal with this issue, each segment must be broken into a smaller number of segments so that each subsequent segment can be mapped into 3D space appropriately. The resolution of this subdivision is dependent on the needs of the construction process. The more the segments are subdivided, the more points will need to be mapped into 3D space, and so the longer the computations and processing will take. Conversely, lower resolutions will produce pathing that does not conform very well to the surface upon which construction should take place. With this issue addressed, straightforward nonplanar construction can be performed on a reasonable surface.

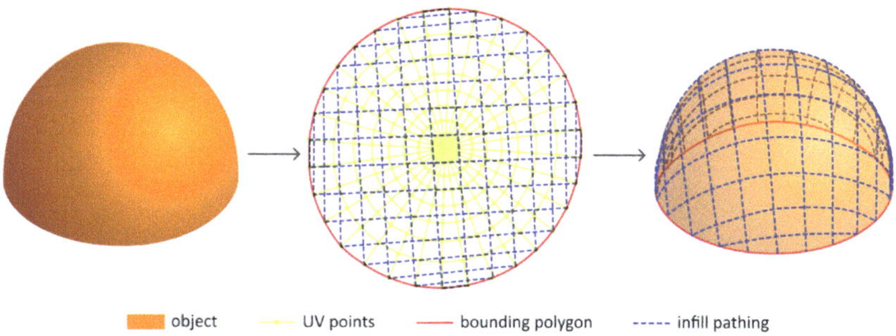

FIG. 15.5 Conformal mapping process for a semisphere (left). The upward facing triangles are conformally mapped to UV space, resulting in a circle (center). The bounding polygon of this circle then has infill pathing (dashed purple lines), as would be generated by a normal slicing process. Finally, the reverse mapping is applied to produce segments on the semisphere (right).

Conclusion

This chapter has introduced extensions to the typical slicing process in the form of off-axis and nonplanar printing. Using off-axis approaches allows construction with next-generation hardware as well as removing the need for supports in some situations. Conformal mapping allows construction directly on nonplanar surfaces. These are two approaches among a variety of pathing solutions and will be built upon in the discussion of 5-axis systems in the next chapter. As new, unique hardware for next-gen systems is developed and construction of ever more complex objects is necessary, the breadth of potential pathing solutions will increase as well. However, more advanced pathing solutions can be developed from the foundation formed by the processes introduced here.

References

[1] P. Chesser, B. Post, R. Lind, A. Roschli, C. Atkins, A. Boulger, P. Mhatre, P. Lloyd, Skybaam large-scale fieldable deposition platform system architecture, in: In 2019 International Solid Freeform Fabrication Symposium. University of Texas at Austin, 2019.

[2] Rotation Matrix, Wikipedia, Wikimedia Foundation, 31 October 2022. https://en.wikipedia.org/wiki/Rotation_matrix.

[3] Quaternion, Wikipedia, Wikimedia Foundation, 16 October 2022. https://en.wikipedia.org/wiki/Quaternion.

[4] X. Gao, S. Loriot, A. Tagliasacchi, Triangulated surface mesh skeletonization, in: CGAL User and Reference Manual. CGAL Editorial Board, 5.5.2 edition, 2023.

[5] L. Saboret, P. Alliez, B. Lévy, M. Rouxel-Labbé, A. Fabri, H. Jain, Triangulated surface mesh parameterization, in: CGAL User and Reference Manual. CGAL Editorial Board, 5.5.2 edition, 2023.

CHAPTER

16

Toolpath considerations for 5-axis systems

Thomas Feldhausen[a], Abby Barnes[a], Alex Roschli[a], Michael Borish[a], Eric MacDonald[a,b], and Peter Wang[a]

[a]Manufacturing Demonstration Facility, Oak Ridge National Laboratory, Knoxville, TN, United States, [b]University of Texas at El Paso, El Paso, TX, United States

Abstract

There have been many studies that show how additive manufacturing (AM) has been able to improve the performance of a component, reduce manufacturing costs, or minimize lead times. As the adoption of AM significantly increased, so did the complexity of designs for AM. Many designs require advanced systems, and in response to this, there is an increasing number of systems now capable of additional degrees of motion. 5-axis systems, which include two rotary axes, are becoming more common in the additive manufacturing community. These systems are capable of depositing material on a nonplanar surface and lend themselves to many industrial applications, further extending the design space for AM. However, computer-aided manufacturing (CAM) for AM has been dominated by 3-axis slicing software that do not typically accomodate 5-axis motion. This chapter introduces 5-axis motion and presents the unique considerations that need to be accounted for when generating toolpath trajectories for 5-axis systems.

Introduction

Additive manufacturing (AM) has enabled the production of objects whose geometries were impossible to produce using traditional techniques. Similarly, 5-axis AM systems can produce objects that are more complex than those made by 3-axis AM systems. While some forms of AM, such as powder bed fusion or material jetting, exclusively use 3-axis motion and slicing techniques, other forms, such as directed energy deposition or material extrusion, commonly leverage 5-axis motion

https://doi.org/10.1016/B978-0-443-15286-3.00020-5

systems. This book has presented the concepts, techniques, and considerations associated with motion and path planning, but has, for the most part, been limited to 3-axis motion. *Chapter 3: Motion of platforms* introduced the nomenclature and some characteristics of 5-axis systems; this chapter will build on those topics by exploring the uses, important considerations for, and recent implementations of advanced path planning for 5-axis motion in AM.

As mentioned in *Chapter 3: Motion of platforms*, 5-axis manufacturing systems commonly include two rotary axes in addition to the three standard Cartesian axes of motion (*X*, *Y*, *Z*). The name of each rotary axis corresponds to the Cartesian axis that the rotation occurs around. As shown in Fig. 16.1, an *A*-axis rotates about the *X*-axis. Similarly, a *B*-axis and *C*-axis rotate about the *Y*-axis and *Z*-axis, respectively. Their direction of positive motion always follows the right-hand-rule; if a person's right thumb is pointed in the direction of an axis, and the other fingers are curled into a fist, the direction that the fingers are curling represents the direction of positive motion.

Although the software used to generate toolpaths for 3-axis systems is often called a slicer, the software used to generate 5-axis toolpaths is often referred to as computer-aided manufacturing (CAM) software. A CAM software is any software used to automate machinery, which always involves the generation of toolpaths and the conversion of toolpaths into machine-readable code. Technically, that is exactly what a slicer does, and a slicer can in fact be considered a CAM software. However, a CAM software often comes with features that expand beyond what a slicer is capable of, particularly through the coded control of the orientation of the printed object and the machinery. CAM software was originally developed for machining applications and has existed for many decades, but advanced CAM software for 5-axis additive toolpath generation is not as refined.

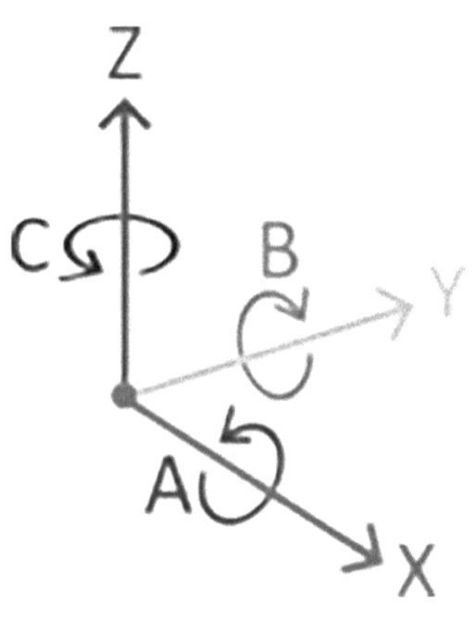

FIG. 16.1 Naming convention for rotational axes of motion.

Applications for 5-axis toolpaths

Analogous to machining, 5-axis systems often provide more degrees of freedom (DOF) than are necessary. Most 5-axis systems in machining are used in a 3+2 configuration. This expression means that most of the motion occurs in the three cartesian axes, while the two rotary axes are used for positioning only. While it may seem like a 3-axis motion solution is sufficient for all positioning, the ability to use precise rotational positioning drastically reduces setup time for complex parts that are traditionally repositioned using complex fixturing. While 5-axis motion is not needed in all AM applications, there are a few cases in which 5-axis motion is advantageous: complex surfaces, decomposed geometries, and discontinuity mitigation. These cases are discussed in the following subsections.

Complex surfaces and surface normality

In AM, the additional axes on 5-axis systems are sometimes necessary when complex surfaces are involved. While complex shapes and surfaces are easily produced when the entire part is additively manufactured, it becomes much more complex during coating and repair operations, where the contour of the part must be precisely followed. In many situations, these operations require the deposition or extrusion nozzle to always remain perpendicular to the object [1]. Unlike the spherical end of a milling cutter that cuts the same no matter the angle of the tool axis, the shape of a circular laser beam, when projected onto a surface, changes as a function of tilt angle relative to the surface. Thus, simultaneous 5-axis motion is used to keep the tool axis vector perpendicular to the surface.

Although 5-axis systems are sometimes necessary for complex surfaces, it's important to limit the amount of 5-axis motion used. Toolpaths over complex surfaces may require many axis reversals that are difficult for an AM machine to achieve because, as mentioned in previous chapters, they can't maintain smooth motion while changing directions at constant speed. They must slow down, but because the extrusion or deposition rate is constant, reducing the speed results in overbuilding and other defects. Therefore, axis reversals are like excess motion and need to be kept at a minimum. As discussed in *Chapter 5: Cross-sectioning*, minimizing this excess motion can be achieved by using graphs and dynamic programming to smooth the surface [2–5]. It's also been demonstrated that advanced control systems can be incorporated in AM systems to provide

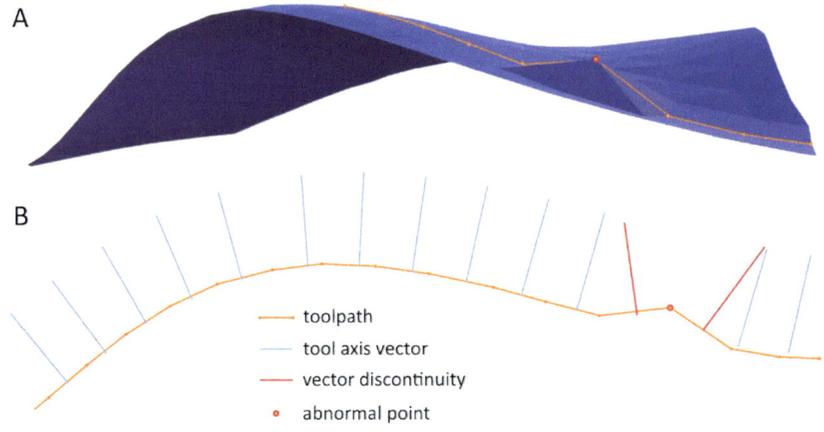

FIG. 16.2 (A) meshed surface with a node placed incorrectly (dot highlighted in red); (B) resulting discontinuity in the tool-axis inclination vectors (red vs blue lines).

feedback to the process to mitigate overbuilding concerns [6], which will be discussed in a later chapter.

It is also important to ensure that all of the normals of complex surfaces are oriented correctly. When repairing components that are worn or damaged, a 3D scanning technique is often used to generate a point cloud, which is an accurate representation of the actual component and was mentioned in *Chapter 4: Geometry data storage*. These operations generally produce a large number of data points that can used to generate a 3D mesh [7]. When these 3D meshes are used to generate toolpath trajectories, using the surface normal is not sufficient, because inaccuracies in the scan will produce misoriented faces in the 3D mesh. Each triangular face that is incorrectly oriented (even the slightest amount) will cause a large discontinuity with the tool axis vector, as shown in Fig. 16.2. This discontinuity can generate a defect in the resulting component, or even damage the motion system if an unwanted collision occurs. Thus, it's common to either smooth the resulting surface or use a 3+2 slicing method to mitigate excess 5-axis motion. More information on repair operations is provided in *Chapter 17: Pathing considerations for hybrid systems*.

Decomposed geometries

In addition to coatings and repairs, 5-axis slicing strategies can also be used to manufacture entire features. In directed energy deposition (DED), discussed in *Chapter 13: Pathing considerations for DED: Arc welding and laser welding*, it's common to decompose the final geometry into individual features that then have distinct toolpaths. For example, starting with a readily

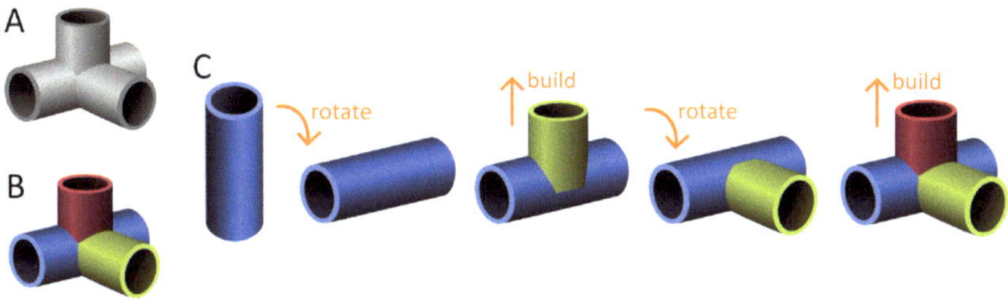

FIG. 16.3 An illustration of 5-axis feature addition; (A) the component to be manufactured; (B) the decomposed geometry; (C) sequential changes in reorientation during the manufacturing process.

available casting or forging and adding features to it can be more economical then using AM to print the entire feature. Hence, this form of AM is commonly associated with 5-axis machines. An example of how a CAD model can be decomposed and reoriented throughout the entire AM process is shown in Fig. 16.3.

Reorientation of a part during construction is also beneficial for the construction of overhangs. As discussed in *Chapter 10: Path modifiers*, many AM processes are limited by the achievable degree of overhung geometry. To expand their capability, support structures are often included into the slicing strategy. However, support structures can be difficult to implement and require costly post-processing to remove [8]. Using a rotary axis to reposition the part eliminates the need for support structures, because deposition can occur in any direction relative to the part [9].

AM discontinuity mitigation

5-Axis motion systems are not always designed with AM in mind and their individual capabilities must be considered during the toolpath generation process. As discussed in the previous section, it's important to maintain a constant speed to mitigate overbuilding due to acceleration/deceleration of the motion system. In addition, there is a potential defect introduced each time the additive process is started or stopped. Thus, the use of continuous and simultaneous motion is very advantageous in additive manufacturing.

A recent development in 5-axis CAM is a toolpath cycle called additive turning. Adapted from 5-axis turning (lathe) operations that are common in machining, this cycle allows entire axisymmetric components to be manufactured with varying wall-thicknesses without any process discontinuities [10] (Fig. 16.4). This cycle generates a three-dimensional spiral. It allows full control over the angle of the tool, called the tool inclination, which is necessary for smooth 5-axis motion. The cycle finds the tool

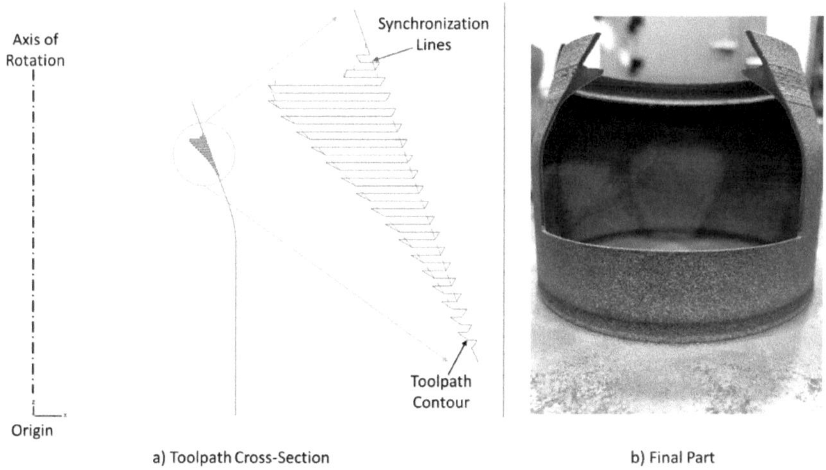

FIG. 16.4 An example of an additive turning toolpath; (A) a schematic of toolpath; (B) the final part produced by the toolpath. *Credit: T. Feldhausen et al., Performance of discontinuity-free components produced by additive turning computer aided manufacturing strategy, J. Mater. Process. Technol. 308 (2022) 117732.*

inclination at all points along the toolpath through a linear interpolation of the tool inclination at various, distinct points. These distinct tool inclinations are often referred to as "synchronization lines" in machine tool (lathe) operations and are generated by the CAM software. The linear interpolation of the synchronization lines provides smooth 5-axis motion throughout the entire build, rather than discrete movements of the tool axis inclination. This is an instance of an AM issue being overcome by a CAM solution, rather than using advanced system controls.

Methods for 5-axis motion

In many cases, 5-axis toolpaths are simply derived from a traditional toolpath by modifying the tool-axis vector. The following subsections do not go in-depth on how these toolpaths are generated, but rather present various approaches to the modification of tool-axis vectors.

Fixed strategy

As previously mentioned, one of the most common methods is a 3+2 approach, hereafter referred to as a "fixed" strategy, where the two rotary axes are fixed in a specific orientation. This provides a constant lean and/or lead angle to the tool axis vector. This is advantageous in situations where machine tool errors may be a concern. Excessive rotary motion makes tool motion more complex, which can result in unwanted or unforeseen issues. Small errors in the representation of the machine tool's center of rotation can be amplified in large parts and produce geometrical errors in the final component. Thus, once again, it's good practice to limit excess motion wherever possible.

Automatic inclination

An important and often overlooked aspect of 5-axis toolpath generation is collision avoidance, which can be done through computer simulations. However, without an accurate representation of the process, it's difficult to account for collisions between the additive equipment and the part being manufactured. Thus, another common 5-axis inclination strategy is to use built-in collision detection, which is included in many CAM software packages. The process works by automatically changing the inclination of the deposition nozzle to remove any potential collisions. This technology is more common in machining CAM software packages but is becoming more commonplace in AM thanks to the commercialization of hybrid manufacturing techniques, which combine both additive and subtractive processes in one system.

Guide curves

For coatings and repair operations where it's required to deposit material into a corner or cavity, it may be useful to use a guide curve to provide the inclination of the tool. Here, an arbitrary curve is drawn in space such that the tool axis vector intersects the curve throughout the entire deposition process. An example of this is shown in Fig. 16.5.

Constant tool inclination

Like many traditional welding strategies, it's also common to include a slight lead or lean angle to the tool axis vector in some applications, such as ball mills where the cutting speed is effectively zero at the tip. For this strategy, a specific lead or lean angle is applied to the tool axis vector. This differs from the fixed strategy, in which both rotaries are in fixed positions, because this strategy typically allows one rotation axis to rotate freely (commonly a *C*-axis) while the angle of the other rotary axis is fixed at the value of the desired lead or lean.

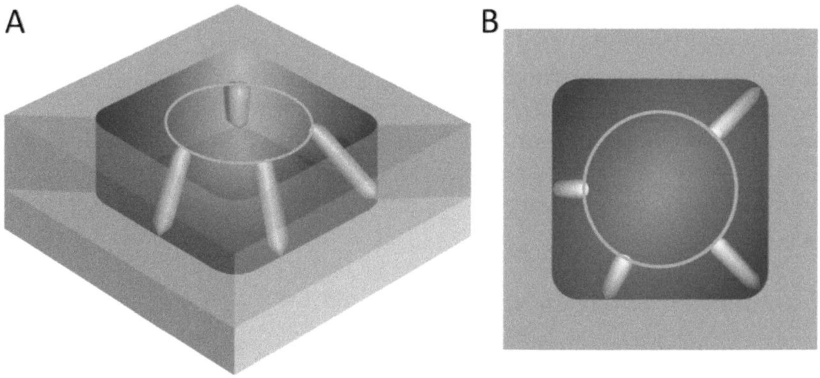

FIG. 16.5 An illustration of guide curves for tool inclination; (A) isometric view; (B) top-down view.

Case study: Slicing approach for large format 5-axis motion

There are too many approaches to, and solutions for, 5-axis toolpathing to cover them all in this book. However, to illustrate the concepts and considerations for 5-axis toolpathing, this section will describe, in detail, one slicing approach that can be used for a fixed, purely additive system. Note that this section builds on the pathing approaches laid out in previous chapters; those pathing approaches and considerations still apply here.

This example will focus on a large format, laser-based DED system. In large format systems, it is not uncommon to find limitations on additional degrees of freedom. Recall that in a fixed (3+2) setup, the +2 is usually provided by manipulation of the build table via tilt and rotation. One common limitation is that the rotation of the table is limited to ± 180 degrees, instead of being unconstrainted.

The general approach to generating a toolpath solution for such a system differs from what has been previously described in this book only by the inclusion of additional information. For instance, cross-sectioning takes place as usual, but the generated toolpath segments are augmented with information about the surface normal. This normal information will be used to continuously tilt the table to keep the object oriented with gravity, which is necessary to produce overhangs without support. Recall that the slicing plane intersects any triangle of the STL at two points, I_0 and I_1, which was covered in *Chapter 5: Cross-sectioning*. Each one of these points lies on the boundary of two faces of the object. Therefore, depending on the ultimate directionality of the closed-loop path that results from the polygon, one of two face normals will be relevant. The relevant normal cannot be determined at this time, as the direction (clockwise or counterclockwise) of the rotation is not known. However, the normal information can be passed along during the pathing computation phases to be determined later. The polygon can then have closed-loop paths computed as usual. It is important to note that, for any path beyond the original polygon, no normals exist. The normals then must be interpolated based on existing geometry. One easy solution for this is to use a nearest neighbors' approach to interpolate the normal based on the original polygon.

Once all the paths have been determined, during optimization, the ordering of the pathing is adjusted. Recall from *Chapter 12: Pathing considerations for extrusion: Pellet, filament, concrete, and thermoset* that one of the optimizations that must happen to closed-loop pathing is adjustments for directionality. With a restricted table rotation, the paths are alternated between counterclockwise and clockwise rotations, depending on the location of the paths relative to the maximum angle of rotation. This means that, once the table reaches its maximum rotation, the direction of the pathing will need to reverse as the table starts spinning in the opposite direction. This alternation between clockwise and counterclockwise orientations is illustrated by the deposition of a simple circle, shown in Fig. 16.6.

With the rotations in place, it is now possible to choose one of two normals that has been passed along since the cross-sectioning step. Based on the directionality of the loop, the chosen normal will be the one that represents the face from the incoming direction. This normal is used to compute a cross product with a normal that represents gravity, which simply points down. The result is the direction the table must tilt to maintain a level build surface. In this way, overhanging structures can be built without support. Examples are shown in Fig. 16.7 [11].

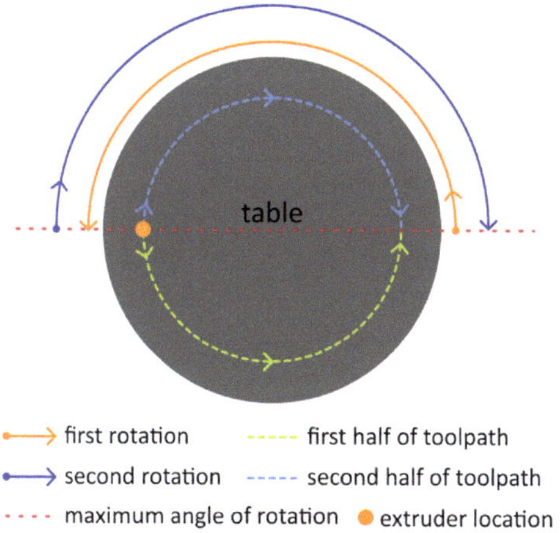

FIG. 16.6 Alternation between counterclockwise and clockwise paths to deposit a circle on a restricted turn table.

FIG. 16.7 Examples of components produced with large format 5-axis motion.

As shown here, the construction approach is very similar to the standard pathing approach but includes the additional layering of information about surface normals and directionality. Naturally, if the system does not have a restriction on the additional degrees of freedom, the approach is made simpler. For example, if the system can rotate unconditionally and this direction is known, the identification and assignment of normal information can happen immediately.

Conclusion

5-axis AM systems are becoming more common in the industrial marketplace. With this trend, it's expected to see an increased demand for 5-axis toolpath generation. This chapter introduced the various areas where the use of 5-axis toolpaths make sense, namely in the manufacture of objects with complex surfaces, decomposed geometries, and discontinuity mitigation. This chapter also briefly discussed the various ways that toolpaths for 5-axis motion can be generated, as well as considerations that need to be accounted for when working with a 5-axis system. The implementation of these advanced toolpath strategies will further broaden the design possibilities and application spaces for additive manufacturing, by enabling the motion required to manufacture the net shape of complex components.

References

[1] D. Plakhotnik, et al., CAM planning for multi-axis laser additive manufacturing considering collisions, CIRP Ann. 68 (1) (2019) 447–450, https://doi.org/10.1016/j.cirp.2019.04.007.

[2] A.P. Pashkevich, A.B. Dolgui, O.A. Chumakov, Multiobjective optimization of robot motion for laser cutting applications, Int. J. Comput. Integr. Manuf. 17 (2) (2004) 171–183.
[3] D. Plakhotnik, B. Lauwers, Graph-based optimization of five-axis machine tool movements by varying tool orientation, Int. J. Adv. Manuf. Technol. 74 (1) (2014) 307–318.
[4] D. Boisselier, S. Sankaré, T. Engel, Improvement of the laser direct metal deposition process in 5-axis configuration, Phys. Proc. 56 (2014) 239–249.
[5] Y. Sun, J. Xu, C. Jin, D. Guo, Smooth tool path generation for 5-axis machining of triangular mesh surface with nonzero genus, Comput. Aid. Des. 79 (2016) 60–74.
[6] B.T. Gibson, B.S. Richardson, T.W. Sundermann, and L.J. Love, "Beyond the toolpath: site-specific melt pool size control enables printing of extra-toolpath geometry in laser wire-based directed energy deposition," Appl. Sci., vol. 9, no. 20, p. 4355, 2019. [Online]. Available: https://www.mdpi.com/2076-3417/9/20/4355.
[7] M.J. Kim, Rapid Digitization and Non-Rigid Registration of Aerospace Blade Geometries for Hybrid Manufacturing Component Repair, Georgia Institute of Technology, 2019.
[8] T. Page, Design for Additive Manufacturing, LAP Lambert Academic Publishing, 2011.
[9] D. Zhao, W. Guo, Shape and performance controlled advanced design for additive manufacturing: a review of slicing and path planning, J. Manuf. Sci. Eng. 142 (1) (2020) 010801.
[10] T. Feldhausen, et al., Performance of discontinuity-free components produced by additive turning computer aided manufacturing strategy, J. Mater. Process. Technol. 308 (2022) 117732.
[11] B.T. Gibson, et al., Controls and process planning strategies for 5-axis laser directed energy deposition of Ti-6Al-4V using an 8-axis industrial robot and rotary motion, Addit. Manuf. 58 (2022) 103048.

CHAPTER

17

Toolpath considerations for hybrid additive manufacturing

Thomas Feldhausen[a], *Abby Barnes*[a], *Alex Roschli*[a], *Michael Borish*[a], *Eric MacDonald*[a,b], *and Peter Wang*[a]

[a]**Manufacturing Demonstration Facility, Oak Ridge National Laboratory, Knoxville, TN, United States,** [b]**University of Texas at El Paso, El Paso, TX, United States**

Abstract

Hybrid additive manufacturing combines both additive and subtractive manufacturing processes to fabricate geometrically complex, dimensionally accurate parts that are difficult to make using either additive or subtractive manufacturing alone. Industrial applications for hybrid additive manufacturing include direct manufacturing (building whole parts), feature addition (adding onto existing parts), and remanufacturing or repair operations. To coordinate the additive and subtractive processes, computer-aided manufacturing (CAM) software must generate distinct toolpaths for each process. Each application requires different considerations and inputs when generating the toolpaths. For example, when building a part with an internal cavity that would otherwise be unmanufacturable due to reach or access limitations, a CAM must determine which layer(s) of deposition are appropriate to machine to create the internal cavity. This chapter explores the area of toolpath generation for hybrid manufacturing by discussing the necessary considerations that are unique to a hybrid process.

Introduction

Recall that, in *Chapter 13: Pathing considerations for DED: Arc welding and laser welding*, directed energy deposition (DED) was defined as a process that additively manufactures metal objects. This additive process offers designers the freedom to make geometrically complex components, to use multiple materials within one object, and to reduce material waste. However, DED struggles with dimensional accuracy and surface finish. On the other hand, machining is a process that is known for its ability

https://doi.org/10.1016/B978-0-443-15286-3.00007-2

to produce geometrically precise components with unprecedented surface finishes, but requires large, solid pieces of material, a lot of which ends up wasted. These processes clearly complement each other, and as a result, manufacturing methods are evolving to integrate DED with a traditional machining method.

That integration is an example of hybrid manufacturing. Hybrid manufacturing leverages the benefits of additive and subtractive manufacturing to overcome the disadvantages associated with each [1]. Importantly, hybrid manufacturing occurs in all forms of AM, not just DED. For example, parts that are manufactured using powder-bed fusion are often transferred to an electric discharge machine for build plate removal, then to a machining center for finishing.

One approach to hybrid manufacturing is to additively manufacture an object using an additive system, then manually transfer the object to a machining system. In this chapter, this approach will be termed a hybrid manufacturing "process." The process is relatively straightforward, but it often requires additional time and equipment to safely and carefully remove the object, transfer it to the machining system, and load it in place, all without damaging the object. Additionally, to ensure accurate machining, the object must be precisely oriented within the machining system, which can be challenging. However, these challenges can be mitigated by integrating the additive and subtractive processes into a single manufacturing system, so that the object can be constructed and machined without the need for relocation [2]. This chapter will refer to that approach as a hybrid manufacturing "system."

There are three main applications of hybrid manufacturing. The first is directly manufacturing components, in which parts are manufactured in their entirety. These operations can either be consecutive, so that the entire additive manufacturing process occurs before machining, or in an interleaved fashion, in which each process occurs multiple times during the manufacturing cycle. This area can be considered a way to replace casting and forging.

The second application area is feature addition, in which a complex part is manufactured by adding features to a casting or forging that is readily available. For example, a commonly available forged cylinder can be modified by hybrid manufacturing to produce an injection molding screw with an integrated cooling channel. This minimizes the amount of material that needs to be printed by leveraging the forgings that are already available in the supply chain.

The final application area is remanufacturing and repair operations. Worn or damaged parts, such as molds and dies for the tooling industry, can be easily repaired using hybrid manufacturing. Furthermore, it's also possible to repurpose components using hybrid manufacturing. Instead of having to manufacture a new tooling set each time a design change occurs, it's possible to repurpose existing tooling simply by adding and removing material in the appropriate areas.

Many hybrid manufacturing systems are being formed by integrating additive machinery into a commercial machining center. Similarly, many commercial computer-aided manufacturing (CAM) software packages now offer AM toolpath solutions along with traditional toolpath strategies for machining. However, generating toolpath solutions for hybrid manufacturing operations is still an extremely manual process that relies on a large amount of input from the programmer [1]. Furthermore, different parts and applications require different considerations. This chapter explores these industrial application areas from a toolpath perspective and highlights the unique aspects of each that must be considered when generating toolpaths for hybrid manufacturing operations.

Directly manufacturing components

Using hybrid manufacturing to directly manufacture components removes the need for casting and forging. The ability to additively manufacture entire components and machine them to meet their design requirements within a single manufacturing system has a lot of advantages [2–4]. However, the coordination of these two distinct processes can be time consuming because it requires an iterative approach, rather than a simultaneous and/or continuous one. This section discusses toolpathing considerations that are important for directly manufacturing components using either consecutive or interleaved additive and subtractive operations.

Consecutive operations

In hybrid manufacturing, the term "consecutive operations" refers to when the entire component is additively manufactured before machining occurs. Whether this manufacturing scheme is conducted in a hybrid manufacturing system or with a hybrid manufacturing process, the same considerations apply. One of the most critical considerations for any hybrid manufacturing scheme is geometrical accuracy. In many cases, AM is a preforming technology, meaning that it is used to make objects that are close to, but larger than, the shape and size of the final, desired part. This is referred to as "near net shape" construction and is geometrically achieved by enlarging the object on a layer-by-layer basis, such that every bounding polygon is enlarged, similarly to the dilations discussed in *Chapter 6: Closed-loop toolpath generation*. After a preform is additively manufactured, it post-processed (machined or otherwise) to meet its original design.

Additively manufacturing preforms saves material because it removes the need for a solid piece of material that is big enough to encapsulate the entire bounding box of the desired part. When using expensive materials, the ability to reduce the amount of material needed to construct a part is a huge advantage. It's common practice in AM to minimize the amount of printed material to cut back on material cost and machining time. However, it's extremely important to ensure that the preform that is additively manufactured fully contains the final part. Toolpathing for AM assumes a perfect outcome, but unfortunately, there are often distortions due to shrinkage and residual stress. As shown in Fig. 17.1, a preform can distort such that the part can no longer be contained inside of it, even if the pathing was correctly calculated for an object that was correctly enlarged.

To combat this geometric distortion, an "allowance" is typically applied to the computer aided design (CAD) model or the generated toolpath. This is the distance that the CAD surface or outermost toolpath segment is offset from the original CAD geometry. The required allowance depends on the geometrical shape, material selection, print orientation, and even the generated toolpath. Thus, this value is often selected via institutional knowledge, or by trial and error. For AM processes with large beads, such as wire-fed DED, there is a large allowance inherent to the process. The outermost segment of the toolpath is often placed directly on the outermost CAD surface. Thus, the printed preform will have an allowance equivalent to the radius of the deposited bead, i.e., half a bead width.

Although an allowance is required for hybrid manufacturing, manufacturing parts with too large of an allowance has its own pitfalls. For one, it can reduce cost savings because of the increased material usage and machining time. For geometrically complex parts that are machined with a ball end-mill and a small infeed, the increased machining time can add up quickly. On the other hand, in situations where a larger diameter

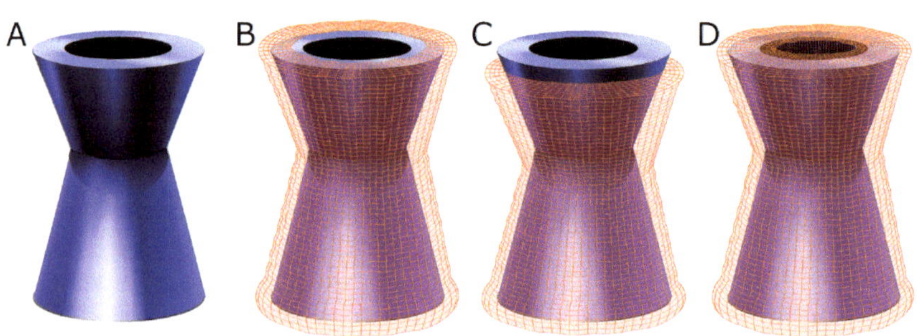

FIG. 17.1 Examples of the suitability of an additively manufactured preform (orange) for the intended part geometry (blue); (A) the desired geometrical shape; (B) unsuitable preform because the inner diameter is too large; (C) unsuitable preform because it is too short; (D) suitable preform because the part is completely contained inside of it.

subtractive tooling can be used (e.g., greater than 6.35 mm), increasing the allowance doesn't affect the machining much. Most machining operations aren't changed if the depth of cut slightly increases.

Increase machining time is not the only risk that arises from a part that is too big, however. There is also a risk of damaging equipment. If too large of a depth of cut is taken, the part that is being machined could be damaged, the cutting tool can break, or the control system can be impaired. This can be mitigated by completing a rough machining operation to remove excess material before doing a finer machining operation for surface finish, but that again increases machine time. Additionally, inspection techniques, such as 3D scanning or touch-probes, need to be used to accurately measure the printed geometry in order to correctly complete the machining operations. Nowadays, many commercial CAM software packages can use inspection data to adaptively control the generated machining toolpath.

Another consideration for consecutive operations is the use of AM to make the machining operation easier. For example, AM can be leveraged to produce preforms that have stiffening, dampening, or fixturing features that be advantageous to the machining operation. For hybrid processes where the part is transferred from the AM machine into a machining center, fixturing an additively manufactured part that has rough edges and curvature is not trivial. Unlike traditional machining operations that start with simple prismatic shapes, customized clamping jaws (known as soft jaws) need to be machined, or clamping features need to be added to the printed preform. Usually, adding clamping features is the fastest and easiest option.

Interleaved operations

One advantage of hybrid manufacturing systems is their ability to interleave multiple processes together, rather than complete one process before moving to the next. Interleaving operations involve the repetition of two steps: additively manufacturing a portion of the final geometry and then machining that constructed portion before moving on to the next portion. This approach has been shown to reduce manufacturing time by 68% and improve material properties, such as ductility [5]. Additionally, interleaving affords easier machining of large parts with high aspect ratios, which are difficult to machine using traditional techniques. For example, to machine the inner surface of a 2-m-long conical shape would require an expensive, custom-made boring bar that is 2 m in length. However, if constructed with interleaved additive and subtractive operations, the conical shape can be printed in multiple sections and each section can be machined using off-the-shelf subtractive tooling.

Interleaved machining also enables the creation of features that are virtually impossible using traditional manufacturing techniques. For example, an enclosed pressure vessel with internal features, such as baffles, must traditionally be constructed as a welded assembly of separate parts. However, using an interleaved hybrid manufacturing approach, the internal features can be machined while the subtractive tooling has reach and access, removing the need to weld separate pieces together. This concept also reduces the need for 90-degree milling spindles, which are often used when reach and access is difficult to achieve. Overall, enabling the machining of features throughout the AM process reduces the complexity of the machining process.

When generating toolpaths for interleaved hybrid operations, the two most important considerations to account for are (1) blending and geometrical mismatch between the sections, and (2) restarting the AM process after machining. In interleaved hybrid manufacturing operations, the cyclic heating and cooling can cause large distortions in the manufactured component. Depending on the shape, these distortions can cause mismatch between the sections. To help mitigate this, as shown in Fig. 17.2, the top portion of the first section is often left untouched to help mitigate concerns of geometrical mismatch. The resulting "ledge" also provides a surface for subsequent material to be deposited on.

Depending on whether a metal wire or metal powder feedstock is used, the ledge may or may not be critical for a successful interleaved strategy. For powder-blown DED, having a surface such as a ledge to deposit on is critical. For laser-wire DED, it's been shown that, depending on the deposited geometry, an interleaved strategy is possible without the ledge [5].

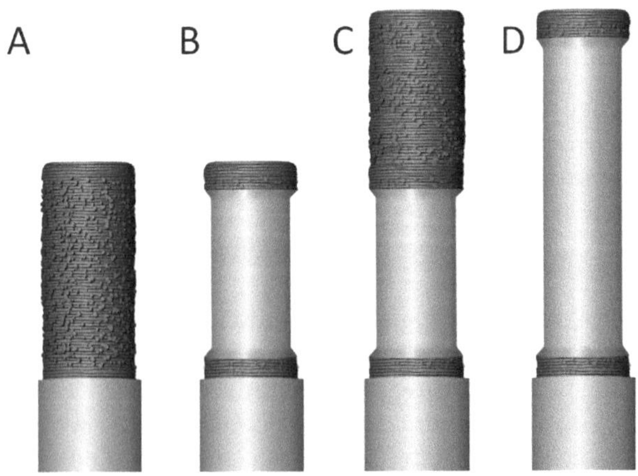

FIG. 17.2 An illustration of interleaved hybrid manufacturing; (A) additively manufactured first section; (B) first section after machining, with an un-machined ledge at the top; (C) additively manufactured second section; (D) second section after machining.

Machining to create a ledge can be easily achieved when using turning (lathe) tools for axisymmetric parts, such as the one showed in Fig. 17.2, but can be more difficult when using a milling operation. The milling of a ledge requires undercut milling tools or an off-axis milling strategy (e.g., use a 5-axis milling machine). However, to avoid this complication, a ledge can also be printed after machining occurs. In that case, the first section is fully machined, and then a ledge is printed as part of the subsequent section by incrementally building layers wider than the machined geometry until the desired width is reached.

For polymer extrusion systems, interleaved operations can be a challenge because of heat management. Each layer is deposited while the previous layer is still hot, which allows the layers to bond together. If the previous layer has cooled, it will not bond as well to the layer being deposited. Interleaving a machining operation allows more time for the layers to cool, which creates a cold joint between sections that is likely to eventually separate. To combat this, the previous layer can be reheated, but the layer-to-layer bond strength can never be fully restored [6].

Feature addition

While AM is most known for the direct manufacture of components and preforms, it is also possible to add features to existing parts. In *Chapter 15: Off-axis and nonplanar slicing*, a method of pathing for printing on top of a curved and/or complex substrate, instead of a flat surface, was discussed. One use of such pathing methods is feature addition. Addition of features is difficult to do with some forms of AM, such as laser powder bed fusion (LPBF), but is easily achievable with DED. In fact, one of the major shortcomings of DED is the long cycle-time, but starting with existing castings and forgings can mitigate that drawback and has been shown to reduce material costs by up to 97% [7]. The difference between a directly manufactured object and the same object manufactured via feature addition is shown in Fig. 17.3.

When starting with existing casting and forgings, the surface of the original object must be appropriately prepared. Machining the surface in question will ensure that the starting stock is properly sized and uniform, and it will also remove any surface contaminants. Depending on the alloy being deposited, the surface may need to be chemically cleaned with alcohol to remove any oils from the cutting fluids. However, recent studies have shown that depositing austenitic stainless steels using a blown-powder DED process without removing the cutting fluid or cleaning the substrate did not have any adverse effects on the hardness or tensile properties [8].

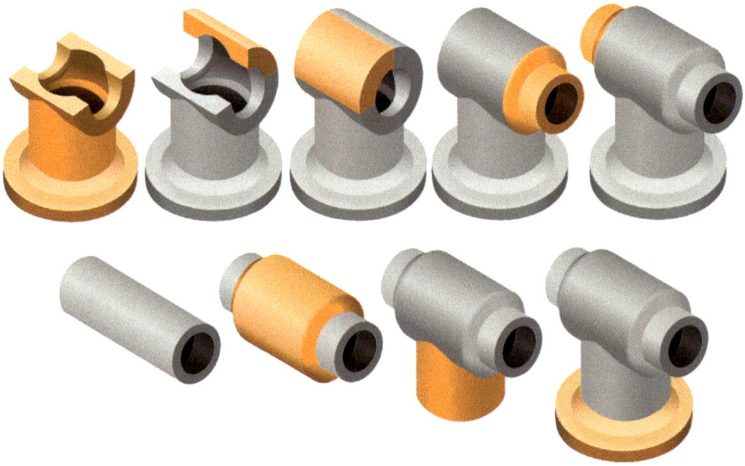

FIG. 17.3 Addition of features to existing geometries; Top row: steps to directly manufacture the entire component; Bottom row: steps to manufacture by adding features to an existing forging. Orange areas indicate the part of the object being deposited in each step.

Another important thing to consider when starting with an manufactured object is that the finished surfaces of the object could become damaged by the construction process. In blown-powder DED, the excess powder can act as abrasive or even stick to the finished surface. In comparison, the wire-fed DED process is cleaner but can have a larger amount of spatter that can land on and bond to surfaces. To combat this, it's common to use welding blankets or aluminum adhesive tape to help protect already finished surfaces. In certain cases, the machining of the starting surface can be delayed until deposition is occurring far enough from the surface that the risk of exposure to these damaging byproducts is minimal.

In this type of hybrid manufacturing, it's common for the material of the substrate to be different than that of the added material. If the two alloys are compatible, a sharp transition zone is acceptable. If the two alloys are not, a functionally graded material (FGM) may be needed. A FGM is produced by blending the two alloys in various ratios, slowly increasing the amount of one material over the other to create a smooth transition. In certain cases, a third alloy may need to be incorporated to further help the metallurgical bonding between the alloys.

Remanufacturing operations

Hybrid manufacturing is particularly important for the repair or re-purposing, collectively called "remanufacturing," of worn or damaged high-value components. Remanufacturing operations typically involve the removal of damaged material, followed by deposition of repair material, and then machining to the final geometry. Traditional remanufacturing has relied on skilled tradesman or a series of disjointed processes. Hybrid manufacturing systems can overcome this paradigm as they enable different process modalities—including inspection, additive, and subtractive manufacturing—on a single machine platform.

CAM strategies for part repair can be classified as "nonadaptive" (rigid) and "adaptive" (nonrigid) path strategies [9,10]. Nonadaptive repair is typically used on parts that experience a lot of wear-and-tear, such as molds and dies, and is performed on surfaces that have been damaged from intense use. In such a case, the worn material is first machined away before material is deposited to build it back up to, or slightly above, the intended geometrical shape. Then, the component is machined to the original CAD specification. These are considered non-adaptive toolpath strategies because the repair is conducted in reference to the initial CAD geometry, rather than to the geometry of the damaged part; the damaged surface is machined away and therefore does not need to be considered during the repair. Importantly, the geometrical mismatch between the initial and target geometry, as well as the registration of the actual component datums, still needs to be recorded.

In comparison, components that become distorted from use, rather than simply worn down, often require an adaptive CAM strategy because the repair is dependent on the current geometry of the component. To correct any deformation, the toolpath must be adapted to the actual, rather than the intended, geometry. The original toolpaths must be adapted to the distorted geometry, which can be measured using probing or laser light scanning [11–13]. Probing strategies leverage existing metrology equipment, usually a contact touch probe, that is available on most machining centers. These probes measure a finite number of points to both determine the shape of a worn component as well as its location within the machine. On the other hand, laser light scanning requires auxiliary equipment, but can generate a larger number of data points for higher resolution mapping [14]. Either way, the generated geometric data must be incorporated into the CAD model or the toolpath trajectory to generate an adaptive toolpath. Otherwise, a geometric mismatch between repaired and nonrepaired sections can occur.

The finishing and blending of repaired components can be difficult due to distortion. The heat input from the additive process produces distortion through thermal expansion and residual stress. Furthermore, these distortions are difficult to quantify as the original surface used for probing and scanning is inaccessible because the surface now has material deposited on it. Geometrical modeling and simulation have been found to be useful to predict the distortion. Another mitigation strategy is to machine the entire functional surface [15].

Hybrid manufacturing also enables a form of modular remanufacturing, in which geometries can be machined away and replaced with a different geometry [16]. An example would be the repurposing of an injection mold to another component in less than 72h [9]. Using this strategy, mid-year design changes in stamping dies in the automotive industry can be incorporated into existing tooling, no matter how drastic the change is. This green manufacturing method allows for more components to be reused, reducing waste by increasing the overall lifecycle of the material [17].

Conclusion

Additive manufacturing enables the direct manufacture of components that are difficult to manufacture using traditional manufacturing techniques. Hybrid manufacturing combines an additive and subtractive process that synergize to overcome the individual limitations of each. The additive and subtractive processes can be combined in a singular machine to form a hybrid system, or parts can be transferred from one system to another in a hybrid process. While the toolpathing strategies previously discussed in this book still apply to hybrid manufacturing, this chapter has presented and discussed circumstances that must be considered when generating toolpath trajectories for hybrid processes and systems. Without these considerations, parts produced by AM will not realize their full potential in the industrial market, because a hybrid approach is necessary to overcome the aspects of AM that limit its usefulness in industrial applications.

References

[1] T. Feldhausen, et al., Review of computer-aided manufacturing (CAM) strategies for hybrid directed energy deposition, Addit. Manuf. (2022) 102900, https://doi.org/10.1016/j.addma.2022.102900.

[2] K. Lorenz, J. Jones, D. Wimpenny, M. Jackson, A review of hybrid manufacturing, Solid Freeform Fabric. Conf. Proc. 53 (2015) 96–108.

[3] J.L. Dávila, P.I. Neto, P.Y. Noritomi, R.T. Coelho, J.V.L. da Silva, Hybrid manufacturing: a review of the synergy between directed energy deposition and subtractive processes, Int. J. Adv. Manuf. Technol. (2020) 1–14.

[4] Z. Zhu, V.G. Dhokia, A. Nassehi, S.T. Newman, A review of hybrid manufacturing processes–state of the art and future perspectives, Int. J. Comput. Integr. Manuf. 26 (7) (2013) 596–615.

[5] T. Feldhausen, N. Raghavan, K. Saleeby, L. Love, T. Kurfess, Mechanical properties and microstructure of 316L stainless steel produced by hybrid manufacturing, J. Mater. Process. Technol. (2020) 116970, https://doi.org/10.1016/j.jmatprotec.2020.116970.

[6] V. Kishore, et al., Infrared preheating to improve interlayer strength of big area additive manufacturing (BAAM) components, Addit. Manuf. 14 (2017) 7–12.
[7] T. Yamazaki, Development of a hybrid multi-tasking machine tool: integration of additive manufacturing technology with CNC machining, Procedia Cirp 42 (2016) 81–86.
[8] R. Kannan, T. Feldhausen, K. Saleeby, P. Nandwana, Effect of humidity of build chamber in hybrid manufacturing systems on part performance☆, Manufacturing Letters 32 (2022) 39–43.
[9] K. Saleeby, T. Feldhausen, L. Love, T. Kurfess, Rapid retooling for emergency response with hybrid manufacturing, Smart Sustain. Manufac. Syst. 4 (3) (2020) 245–249, https://doi.org/10.1520/SSMS20200050.
[10] S. Kaierle, et al., Single-crystal turbine blade tip repair by laser cladding and remelting, CIRP J. Manuf. Sci. Technol. 19 (2017) 196–199.
[11] M. Praniewicz, T. Kurfess, C. Saldana, Adaptive geometry transformation and repair for hybrid manufacturing, Proc. Manuf. 26 (2018) 228–236.
[12] X. Zhang, W. Cui, W. Li, F. Liou, A hybrid process integrating reverse engineering, pre-repair processing, additive manufacturing, and material testing for component remanufacturing, Materials 12 (12) (2019) 1961.
[13] M. Praniewicz, T. Kurfess, C. Saldana, An adaptive geometry transformation and repair method for hybrid manufacturing, J. Manuf. Sci. Eng. 141 (1) (2019).
[14] M.J. Kim, Rapid Digitization and Non-Rigid Registration of Aerospace Blade Geometries for Hybrid Manufacturing Component Repair, Georgia Institute of Technology, 2019.
[15] K. Eiamsa-Ard, H.J. Nair, L. Ren, J. Ruan, T. Sparks, F. Liou, Part Repair using a Hybrid Manufacturing System, 2005.
[16] J.K. Nagel, F.W. Liou, Designing a modular rapid manufacturing process, J. Manuf. Sci. Eng. 132 (6) (2010).
[17] S.-K. Rittinghaus, et al., Direct energy deposition of TiAl for hybrid manufacturing and repair of turbine blades, Materials 13 (19) (2020) 4392.

CHAPTER

18

The g-code file

Alex Roschli[a], *Michael Borish*[a], *Thomas Feldhausen*[a], *Abby Barnes*[a], *Peter Wang*[a], *and Eric MacDonald*[a,b]

[a]**Manufacturing Demonstration Facility, Oak Ridge National Laboratory, Knoxville, TN, United States,** [b]**University of Texas at El Paso, El Paso, TX, United States**

Abstract

The slicing process for additive manufacturing (AM) involves many steps, including slicing the object to be printed and fitting toolpaths to each resulting layer. The toolpaths define the trajectories that the machine must follow during printing to construct the object. Once the slicing is complete and all toolpaths have been found and properly planned, this data must be transferred to the 3D printer in a way the printer can use to construct the object. This is achieved by exporting a file, usually a g-code file, from the slicing software. The g-code file is a text-based file that contains instructions for the machine, controlling the system and defining the motions for each toolpath. The file, which often contains thousands of lines of code, is mostly composed of numbers that command positions for axis motions. This chapter will outline the contents of a g-code file, discuss various commands and their implementation, then provide example codes and files for object construction.

Introduction

The preceding chapters of this book have covered how to take an object designed through CAD (computer aided design) and export it as a 3D mesh, slice it into layers, then generate toolpaths for each layer. However, these steps haven't discussed how to go from the planned toolpaths to a data output that a machine can use to print the part. What's needed is a g-code file, which is numerical code that contains instructions that the 3D printer can execute to construct the object. The g-code file includes motion commands and input/output (I/O) commands to control the various systems of the 3D printer. This text-based file can be exported from the slicing program and transferred to the 3D printer to begin object construction.

https://doi.org/10.1016/B978-0-443-15286-3.00002-3

What is g-gode?

G-code is the name given to the programming language used for a CNC (computer numerically controlled) machine. G-code was created to move the machine tool industry from NC (numerically controlled) machines, which were operated with punch tape, to CNC machines, which are operated by computers [1–3]. There are many different types of CNC machines, including mills, lathes, punch presses, and 3D printers. While the functionality of each of these machines varies, the process for communicating with and providing instructions to them is the same: g-code. A g-code file is a text file containing g-code and m-code instructions, or commands, which the machine reads to execute a specific activity, such as extruding material or spinning a tool. G-codes are used to define operational modes or initiate motion sequences, while m-codes are used for enabling/disabling systems and communicating input/output (I/O) commands, such as setting the extruder temperature.

G-code is most commonly stored as a text file with the *.gcode* extension. Other file extensions used for g-code include *.nc* and *.mpf*. G-code files are saved in plain text, without encryption, meaning they can be opened by any text editor from Notepad to Microsoft Word or Google Docs. ASCII (American Standard Code for Information Interchange) character encoding is used for g-code files, which makes them easy for a person to read and parse the commands. Two example g-code files, which each show the instructions to print three layers of a cube shape, can be seen in Appendix B.

G-code syntaxes

A lot of the format and structure of g-code has been standardized, but there is also plenty of variation among systems and manufacturers. Each system uses a different version of g-code and may have its own commands, format, and structure. These versions are often referred to as syntaxes, or flavors. The need for different syntaxes comes from the variation in machine functionality. For example, filament-based thermoplastic systems use positional control and must provide coordinated motion between the gantry and filament extruder. This is done by adding an additional parameter, the E parameter, to each motion command to denote what length of filament to be extruded during the motion. The calculation for the length of the filament, the value of which immediately follows the E parameter, can be found in *Chapter 12: Pathing considerations for extrusion: Pellet, filament, concrete, and thermoset*. Meanwhile, pellet-based thermoplastic systems use velocity to control the extruder. This is implemented with an M-code to turn the extruder on/off. For example, the line `M3 S400` is used to turn the extruder on at 400RPM.

Other differences can be found in the I/O's that a specific machine has as well as any subsystems that may need to be enabled or disabled. For example, the Cincinnati BAAM system uses a tamper during printing, which is initiated by the M64 command. Another example is the development of a machine-specific, preprogrammed motion to a known location, which can be implemented through a specific m-code that then calls a motion, or sequence of motions, to reposition the system. This functionality is used for parking the system at the end of a print, moving to a designated location for purging and/or initiating a purge, or moving to a location to change extruders or tool heads for a system that contains multiple tools. These commands that are specific to the functionality of a machine are not standardized and thus create the need for individual machine syntaxes. To demonstrate these differences, the two sample g-code files included in Appendix B use different syntaxes: the open-source RepRap syntax and the closed-source Cincinnati BAAM syntax.

Motion commands

Commands contained in a g-code file can be grouped into two categories: motion commands and nonmotion commands. Motion for AM equipment comes in three primary varieties: linear, arc, and spline. All three of these motions can be used during extrusion toolpathing, but not all systems and syntaxes support all the motion types. In particular, arcs and splines are not always supported due to the amount of computation required to accurately interpolating the pathing. Therefore, the pathing for moving between points, and the number of axes engaged to create the motion, can significantly vary.

As previously mentioned, motion commands are implemented with g-codes. While an m-code can be used to initiate motion, the m-code is typically used to call a program that is using g-codes to execute the motion. The remainder of this section will focus on common g-codes used to create motion and how they can be implemented.

G0 and G1—Linear motion

A linear motion is a simple move wherein the system moves directly from one location to another using at least one of the three X, Y, and Z axes. The command, which can be a G0 or G1, contains the location of the system's target end point. The difference between G0 and G1 is how velocity is defined. G0 commands don't include a velocity value. This is because

G0 is referred to as a rapid move. The velocity for a rapid move is predefined by the machine and applied to every G0 command. G0 is most commonly used for travel moves, when no extrusion is needed. On the other hand, each G1 command can include a velocity (also known as feedrate) value, denoted with the F parameter, to define the speed for the move. If no F parameter is included, the last commanded F value is used. It is up to the system's controller to calculate the acceleration and deceleration for G0 and G1 motions in order to complete the pathing such that there is no overshoot on the velocity or on the position of the endpoint. If the end effector goes beyond the end point, an inaccurate geometry will be printed.

An example g-code line for a G0 motion is `G0 X10 Y0`. In this command, the *X* and *Y* parameters are defined such that the end effector is commanded to go to the *XY* location (10,0) using rapid motion. The same motion can be commanded with a G1, `G1 F100 X10 Y0`. This G1 line adds the F parameter, to denote a speed of 100. Similarly, a *Z* component can be included, such as in the command `G0 X10 Y0 Z5`, which moves the end effector to the *XYZ* location (10,0,5). The distance units for each parameter vary widely among systems, with millimeters being most common.

Certain machine configurations can have two different elements that both move within the same coordinate axis, as mentioned in *Chapter 3: Motion of platforms*. To differentiate their commands, two different parameters are used to indicate the Z coordinate, one for each element. For example, a Cincinnati BAAM machine has Z motion for the extruder and for the build table. The extruder Z motion is commanded via the Z parameter and the table *Z* motion is commanded via the *W* parameter. Both axes can move simultaneously, e.g., `G0 X10 Y0 Z10 W5`, though careful consideration must be taken to avoid collisions.

If only one of the three axes is commanded for the linear motion, such as the line `G0 X100`, only the specified axis moves. When two or more of the axes are commanded within the same motion, the velocity is calculated as a vector such that the motion of each axis completes at the same time. This calculation, called coordinated motion, is done by the machine itself. The slicing program only outputs the absolute velocity that the machine must match. Fig. 18.1 shows an end effector moving from the origin, (0,0), to the location (5, 10) both with and without coordinated motion. The example without coordinated motion illustrates that XY motion is done with a slope of 1 until one axis (in this case, *X*) reaches its commanded location. Then, motion is isolated to the axis that has yet to reach its destination, which ultimately creates a path with two moves. The coordinated motion example shows just a single move.

Filament extrusion works with coordinated motion by adding an *E* parameter to the G1 command. As was previously mentioned, the value of the *E* parameter defines the amount (length) of filament to be passed through the extruder during the motion. The system must calculate a

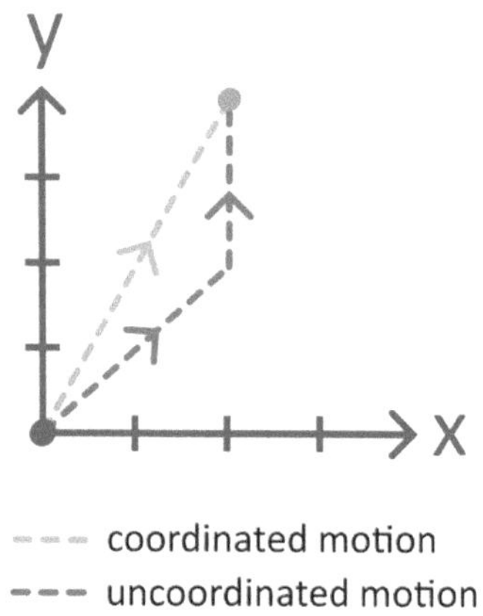

FIG. 18.1 Traveling with and without coordinated motion determines the number of discrete moves.

constant speed for the filament extrusion such that extrusion finishes when the end effector motion finishes. An example of an *E* command in a G1 line: `G1 X3.5 Y8 E1.2`. This line commands the system to move to (3.5, 8) and extrude filament to a length of 1.2. The units of filament distance, just like the *XYZ* parameters, can vary based on machine and syntax, though millimeters is the most common.

An additional parameter, *S*, can be used for laser systems. The value of the *S* parameter defines what laser power to use during the move. The most common unit for the *S* parameter is watts.

G2 and G3—Arc motion

Arc motions are more complex than linear motions because a minimum of two axes must be engaged to follow the path of a circular arc. The command is either a G2 or G3, and both can be defined in two different ways. One notation uses a center point and end point, along with the option to add a velocity. An example is the line `G2 X10 Y10 I5 J3`, where *X* and *Y* define the arc end point, and *I* and *J* define the *X* and *Y* location of the center point. The *I* and *J* values are commanded relative to the current position of the end effector, not in absolute coordinates like *X* and *Y* (the next section elaborates on relative vs absolute coordinates in the G90 and G91 subsection). The other notation for arcs uses the radius rather than the center point. An example is the line `G2 X10 Y10 R5`, where *X* and *Y* define the arc end point and R defines the arc radius. Given just a start point, an end point, and a center point or radius, there are two possible solutions for an arc: one clockwise and one counterclockwise. This is why arc commands can be G2 or G3. G2 is used for clockwise arc motion, and G3 is used for counterclockwise arc motion.

Both G2 and G3 use the same structure for defining the end point as well as the center point or radius. Interpolation of the arc motion is done by the machine controller such that a constant velocity is maintained during the entire motion. Like linear motions, the feedrate can be commanded with the arc by adding the *F* parameter, or the last commanded *F* parameter will be used. This interpolation can also accommodate a third position, *Z*. For the center point arc notation, a *Z* component is added to the end-point location, and a *K* value is used to define the relative *Z* component of the center point. An example can be seen in this line for a counterclockwise arc: `G3 X10 Y10 Z4 I5 J3 K2`. For the radius notation, only a *Z* component is added to the end point. An example can be seen in this line for a clockwise arc: `G2 X10 Y10 Z4 R5`.

Like linear motions, filament systems can add an *E* parameter to the arc motion to define the amount of filament to be extruded during the motion. This *E* parameter is typically tacked on to the end of the line. An example of an arc with commands for velocity and filament length is `G3 F100 X10 Y10 Z4 I5 J3 K2 E12`. Also like linear motions, laser systems can add an *S* to define the laser power.

G5—Spline motion

Spline motions are the most complex, but most flexible, of the three motions. A spline can be fit to a more complex curvature than a circular arc with a constant radius because the spline can use a cubic or quadratic function to model the geometry (note that this section will only focus on writing cubic splines). However, a spline path can only move in two axes at one time, typically *X* and *Y*.

The G5 command is used for spline motions. Additional parameters include *I, J, P, Q, X,* and *Y*. *I* and *J* are used to define the *X* and *Y* location, relative to the start location, of the first control point. *P* and *Q* are used to define the *X* and *Y* location, relative to the end location, of the second control point. *X* and *Y* define the end point. An example can be seen in the line `G5 I1 J2 P4 Q4 X2 Y2`. Additional spline commands can immediately follow the first spline command. When using multiple splines, the *I* and *J* parameters can be omitted from the subsequent splines if the end direction of the previous spline is to be used as the start direction for the current spline. Because of their complexity, very few systems support spline commands.

Like arcs and linear motions, an *F* parameter can be added to define the feedrate, or the last commanded *F* value will be used. *E* can also be added to define filament length to be extruded during the motion. *S* can be added for laser systems to define a laser power.

G28

The G28 command is a motion command for initiating a home sequence. This motion is typically implemented as a G1 command to a predefined "home" location. The home location is defined by the machine controller but is often at the maximum or minimum value of each axis. For example, some systems will home to (0, 0, 0). The G28 command for those systems simply executes one G1 or G0 command to go to the home location, such as `G0 X0 Y0 Z0`. However, some systems must break the home motion into two commands to safely move the *Z* away from the workpiece before traveling in *X* and *Y* to the home location. In that case, the G28 command would execute `G0 Z0` first and then execute `G0 X0 Y0`.

Nonmotion commands

The second type of command contained within a g-code file is the nonmotion command, such as a dwell or a mode definition. Nonmotion commands can be implemented with g-code or m-code, depending on the specific functionality. This section will discuss some common g-code and m-code commands used for AM.

G4

The G4 command is used to insert a pause, or dwell, to the g-code program that stops all motion or additional execution of the g-code program until the pause is complete. The G4 command must be accompanied by an additional parameter to define the pause duration. Most syntaxes use the *P* character followed by a number, in milliseconds, to define the duration. For example, the line `G4 P1000` would pause for 1000 ms. Some machines and syntaxes will use seconds as the default time unit instead of milliseconds.

G54–G59

Commands G54–G59 are commonly used for work offsets, which are used to tell a printer where the origin of an independent coordinate system is relative to the printer's home location. The Cartesian values associated with a G54 command correspond to the distance between the independent origin and the home location. For large printers or CNC systems that may have various parts produced in different locations, these work offsets can be used to easily keep track of the various coordinate systems used to build each part.

G90 and G91

The G90 command is a mode-setting command that tells a machine to read all coordinates as absolute positions. This means that every commanded axis position, e.g., *XYZ*, is executed using absolute position values. These absolute position values are defined based on the workspace origin, meaning that no matter where the end effector is within the workspace, the location of the absolute position command is unchanged. This is in opposition to relative coordinate mode (also known as incremental mode), which commands coordinate values relative to the current position of the end effector. The location of the relative coordinate command is dependent on the current location of the end effector. Relative coordinate mode is initiated with the G91 command. Both G90 and G91 need no additional parameters for execution. An example of a series of G1 motion commands with the resulting absolute versus relative positions can be seen in Table 18.1. Note that the end effector starts at the workspace origin.

G92

The G92 command is used to set the current position or command a coordinate offset. For filament AM systems, the G92 command is often accompanied by an *E* parameter with value zero, i.e., `G92 E0`. This resets the filament axis back to zero for both relative and absolute positioning. If absolute positioning is used without the `E0` command, the *E* value for the filament would continually increase throughout the whole print and could become a very large number. Therefore, it is common to reset the *E* value to zero at the start of a new toolpath.

TABLE 18.1 A series of motion commands and the resultant position for absolute and relative coordinate modes.

Command	Absolute position mode location (*X*, *Y*)	Relative position mode location (*X*, *Y*)
G1 X10 Y10	(10, 10)	(10, 10)
G1 X10 Y10	(10, 10)	(20, 20)
G1 X5 Y0	(5, 0)	(25, 20)
G1 X0 Y0	(0,0)	(25, 20)
G1 X-10 Y5	(−10, 5)	(15, 25)

In this example, the end effector begins at the workspace origin (0,0), then executes each command sequentially.

The G92 command can also be used to modify or update a work coordinate system, such as G54. Using the command `G92 X5 Y0 Z0` would modify the active coordinate system (e.g., G54) such that the current location of the extruder in the global frame is set as the point (5,0,0) within the frame of the specific coordinate system. This is useful when printing multiples of the same part, because the g-code for one part can be repeated by using the G92 command to offset the extruder to a different location on the print bed.

M0

The M0 command is a program stop command. When g-code execution encounters an M0 command, the system comes to a stop and waits for the machine operator to acknowledge the M0 and continue operation of the program. Continuation is usually done with a "Continue" or "Cycle Start" button that the operator must use. The M0 command can be implemented after a step of the printing process where the operator needs to inspect the print before continuing, such as purging the extruder at the start of the print. The M0 allows the operator to ensure the material flow has reached a steady state and is ready to begin the print. No additional parameters are needed with the M0 command.

M1

The M1 command is an optional program stop. The is functionally identical to the M0 command, but the M1 is only executed if "optional stop mode" is enabled by the machine operator. A slicing program will commonly insert an M1 before the start of every layer. This allows the machine operator to safely and easily stop the machine after a layer finishes printing by enabling optional stop mode if they see an issue with the print that doesn't require stopping the machine immediately. No additional parameters are needed with the M1 command.

M3 and M4

The M3 command is known as the "spindle on" command because of its origin in CNC milling, where the cutting head is known as the spindle. For AM purposes, the "spindle" term is still used, but is synonymous with extruder. For pump and screw-based extruders, the M3 command is used to turn on the extruder at a constant rotation speed. An S parameter must accompany an M3 command and is used to define the rotational speed, typically in units of RPM. For example, the line `M3 S400` is used to turn the extruder on at 400RPM. The M3 command executes a clockwise rotation, while an M4 command executes a counterclockwise rotation. Most AM extruders spin only one direction, and therefore only use the M3 command.

M5

The M5 command is used to stop or disable the extruder that was enabled with an M3 (or M4) command. The M5 is issued at the end of each toolpath to stop the flow of material, as well as at the end of the program to ensure the extruder is properly shut off. No additional parameters are needed with the M5 command.

M30 or M2

The M30 command is used to mark the end of the program. Once the machine encounters an M30 command, no additional commands are parsed and executed by the program. Some syntaxes and machines will have M30 as the very last line of code in the g-code file, and others will use the space after the M30, called the footer, to input comments about the slicing settings and printing parameters. While M30 is the most common m-code command used to signify the end of the program, some syntaxes use M2 to mark the end of file. No additional parameters are needed with the M30 or M2 command.

M98

There are times during the execution of a g-code file where the same set of lines, or block, of g-code needs to be executed multiple times. To minimize the length of the file, this block of g-code can be defined as a subroutine, which is then called each time it needs to be executed. The M98 command is used to call a specific subroutine, which stops the program from executing the next line of code, moving instead to the start of the subroutine. Then, all of the lines in the subroutine are executed before returning to the next line of the main program.

The M98 must be accompanied by a parameter, *P*, to define what subroutine is being called. For example, the line `M98 P500` is used to call subroutine 500, which must be defined within the g-code file that is currently executing. The definition of a subroutine is indicated by the letter "O" followed by the number of the subroutine. So, for the previous example, the g-code file would need the string `O500` somewhere within it, which would tell the machine where subroutine 500 starts.

M99

The M99 command is used to mark the end of a subroutine, such that every subroutine starts with `O` and ends with `M99`. When an M99 is executed, the program returns to the M98 that initiated the subroutine, then continues executing the lines that follow the M98. No additional parameters are needed with the M99 command.

M104

The M104 command is used to set the extruder temperature. An additional parameter, *S*, is needed to define the temperature. For example, the line `M104 S230` is used to set the extruder temperature to 230 degrees. Temperature units are typically in Celsius, though some systems can use Fahrenheit. This command is most commonly used by filament systems for extruders that have a single heat zone. However, some machines and syntaxes have adapted the M104 command to read multiple parameters to define additional heat zones on multizone extruders. The M104 command is often placed at the beginning of the program to define the temperature for printing, and can also be placed at the end, typically with an *S* value of zero, to turn off the extruder heat once printing has completed.

M109

A controller's execution of the previously discussed M104 command typically takes milliseconds, and then the next line of code is executed. However, this doesn't leave time for the extruder to reach temperature because the heating process typically takes minutes. If the system gets to a line of code that commands extrusion, but the temperature has not been met, not only would the resulting print probably fail, but the attempted extrusion of cold material could damage elements of the system. This is where the M109 command is helpful. M109 functions the exact same as M104 to set a temperature with the *S* parameter, but it then executes a wait command until the desired temperature is reached. For example, `M109 S250` is used to set the extruder temperature to 250 degrees and then wait until the extruder temperature reaches 250 degrees before continuing program execution. Like the M104, temperature units are typically in celsius, but Fahrenheit can be used by some systems.

Comments

In addition to motion and nonmotion commands, g-code files can also contain comments. Most g-code syntaxes, and systems that run g-code, provide support for text-based comments within the g-code file. Comments are human-readable lines, or strings, of text that are meant to convey a message to an operator looking at the file, rather than to the machine reading the file. Comments are often used to provide setup instructions or to define what is happening in a line of g-code. For AM, this can include everything from defining a nozzle diameter for the operator to install before starting a print, to a comment that signifies the end of a layer or start of a new layer. Comments aren't meant to be read and executed by the machine during printing, but not all systems are capable of ignoring comments and not all syntaxes provide support for comments. As such, those systems and syntaxes must use g-code files that only contain commands to be executed by the controller.

Comments are typically denoted in one of two ways: semicolons and parentheses. The semicolon character, ";", has no function in a g-code file, so it can be used to denote the beginning of a comment, known as a comment delimiter, such that any text following the semicolon, within the same line, is considered a comment and is not executed. For example, the g-code line `G1 X100; PERIMETER` uses the semicolon to begin the comment, which is the string `PERIMETER`. There is no requirement for the semicolon to be preceded or followed by a space or other character. The semicolon can immediately follow a position or command, such as in the line `G1 X100; PERIMETER` where the spaces around the semicolon have been removed.

Parentheses, like the semicolon, are also not typically used in a g-code file and therefore can be used as a comment delimiter to mark the location of a comment. However, if an open parenthesis character, "(", is used to start a comment, then the closed parenthesis character, ")", is required to mark the end of the comment. An example can be seen in the line `G1 X100 (PERIMETER)` where `PERIMETER` is the comment string. Again, a comment denoted by parentheses generally cannot span multiple lines and nothing following the comment on the same line is executed by the program, although there are exceptions to those rules. The parentheses, like the semicolon, do not have to be preceded or followed by any character. The line `G1 X100(PERIMETER)` provides the same functionality as the previous line that had a space before the open parenthesis.

Writing g-code from toolpaths

Writing g-code is the last step of the slicing and toolpath generation process. Most slicing programs only work with linear motions, which can be used for all three of the main toolpath types: closed-loop, open-loop, and space-filling. These linear motions are planned by storing the start and end points of each move in an array—one array per toolpath, and multiple arrays (or an array of arrays) per layer. A g-code writer is used to step through the layers of the entire build, writing g-code that specifies the start of a layer and then writing the appropriate g-code for all the arrays of each toolpath in the layer. Each linear move of the toolpath array can be written with a G1 command.

The array for a toolpath can store more data about each move than just the start/end point. Other data might include layer height, bead width, extruder speed (RPM), filament extrusion length, print speed, and material. Some of these other parameters, such as extrusion speed, are commanded at the start of the toolpath rather than with each linear motion. After all the g-code for the toolpath is written, an additional command must be added to disable the extruder, which prevents extrusion during travels. The following code snippet is for a thermoplastic pellet extrusion system and shows one full toolpath, including the travel to the start of the toolpath.

```
G1 F1500 X0.0000 Y0.0000 Z4.9987 ;TRAVEL LIFT Z
G1 F18000 X509.4320 Y284.9390 Z8.1991 ;TRAVEL
G1 F1500 X509.4320 Y284.9390 Z3.2004 ;TRAVEL LOWER Z
M3 S175 ;TURN EXTRUDER ON
G1 F1920 X708.4750 Y284.3090 ;PERIMETER
G1 X808.5420 Y456.3690 ;PERIMETER
G1 X709.5670 Y629.0620 ;PERIMETER
G1 X510.5240 Y629.6910 ;PERIMETER
G1 X410.4560 Y457.6310 ;PERIMETER
G1 X505.4530 Y291.8810 ;PERIMETER
M3 S50 ;UPDATE EXTRUDER RPM
G1 F1140 X509.4320 Y284.9390 ;PERIMETER SLOW DOWN
M5 ;TURN EXTRUDER OFF
G1 F6000 X529.4320 Y284.8760 ;PERIMETER FORWARD TIP WIPE
```

Note that this code snippet used four points of precision, or four numbers following the decimal for each value, to ensure accuracy. Also, to maximize the human readability of this example code, every line contains a comment to define what is happening within that line of code.

G-code for 5-axis systems

AM systems with 5 axes of motion, discussed in *Chapter 16: Pathing considerations for 5-axis systems*, have some unique aspects to their g-code that are a reflection of the additional axes of motion. Like the g-code files for 3-axis systems, *X*, *Y*, and *Z* coordinates are assigned to each vertex for a 5-axis system. However, vertices for 5-axis systems are also assigned with a tool inclination vector, commonly denoted as $<I, J, K>$.

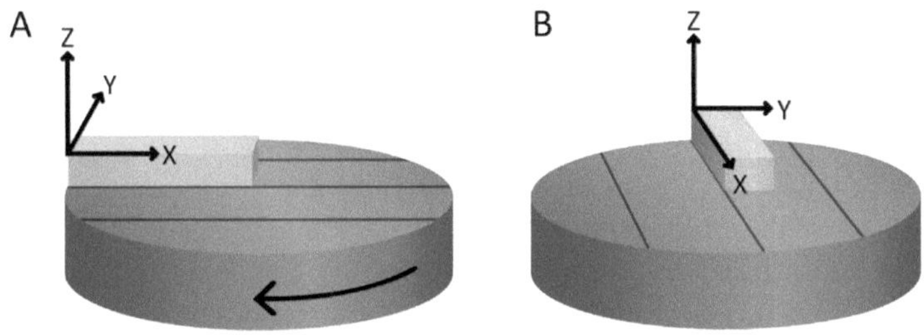

FIG. 18.2 The *X*, *Y* and *Z* axes in the (A) initial state and (B) after table rotation when using Tool-Center-Point-Control (TCPC).

Like the orientation vectors discussed in *Chapter 2: Robot kinematics*, these vectors tell the machine how the tool should be orientated in space as it approaches the vertex.

Another difference is that 5-axis g-code coordinates are usually given relative to the part, rather than the workspace. Older 5-axis g-code files did not have this difference, but were machine-specific; a 5-axis g-code file for one machine could not have been used on a different machine, even if it was the same model. This is because coordinates that are relative to the machine's workspace require an accurate vector representation of the center of rotation of the rotary axis. The representation will be different for each individual machine, because of the inherent differences between their respective rotary axes. The vector representation must exactly match the actual rotary axis of the machine being used. Even a slight offset of the vector representing the rotary axis from the actual rotary axis can be detrimental to the final geometrical accuracy and could potentially damage the machine.

Tool-center-point-control (TCPC) was developed to allow generalization of 5-axis g-code. For modern CNC machines, TCPC enables the system to keep everything in part-specific coordinates, rather than overall machine coordinates. Then, because the CNC controller accounts for where the center of rotation is for each axis, the part can be rotated anywhere in space without affecting the processing of the g-code, as illustrated in Fig. 18.2. This means that the same g-code file can be run on multiple machines (barring any required syntax changes) and is why coordinates for vertices in the g-code for 5-axis systems are defined relative to the part, not the machine's workspace.

Conclusion

G-code is a programming language used to write the toolpath instructions that are read and executed by a 3D printer to construct an object. The slicing process exports this g-code using a g-code writer that translates the toolpath information, such as speed and position, to output commands such as G1, for a linear motion. The g-code writer will also insert proper commands for travels, setting modes, setting temperatures, and other functionalities specific to the machine and process. The output file, called a g-code file, is a text file that contains the g-code instructions, typically with the file extension .gcode. G-code files and formats can vary among machines and processes, with each variety of g-code called a syntax. Once the slicing process is complete, and the g-code file has been exported, the 3D printer can begin construction of the object. Sample g-code files can be seen in Appendix B.

References

[1] G.L. Haggen, A History of Computer Numerical Control, 1988.
[2] G. Younkin, E. Hesla, Origin of numerical control [history], IEEE Ind. Appl. Mag. 14 (5) (2008) 10–12.
[3] S.F. Krar, A. Gill, P. Smid, Computer Numerical Control Simplified, Industrial Press Inc., 2001.

CHAPTER

19

Closing the Loop

Michael Borish[a], *Alex Roschli*[a], *Abby Barnes*[a], *Eric MacDonald*[a,b], *Peter Wang*[a], *and Thomas Feldhausen*[a]

[a]Manufacturing Demonstration Facility, Oak Ridge National Laboratory, Knoxville, TN, United States, [b]University of Texas at El Paso, El Paso, TX, United States

Abstract

Like all research, the concepts of exactly what a slicer should accomplish, and how, continue to evolve. The development of next-generation slicing is already underway, and this chapter introduces several related topics. This includes a new slicing paradigm driven by network connectivity, called "slice on demand." This paradigm enables additional capabilities such as closing the loop with sensor feedback, experimental system support, and connectivity to simulation and visualization software, all of which are also discussed. Each of these concepts represents future opportunities for slicing software and supports the growing expectations of additive manufacturing technology.

Introduction

The previous chapters of this book presented a generalized approach to slicing and path planning, which represents the current state of the art. However, there have also been many allusions to upcoming, next-generation hardware or the limitations of the current slicing process. This chapter will touch on several of these issues and discuss how they are addressed by next-generation slicing. In short, this chapter will answer questions about some of the cutting-edge research currently being undertaken with respect to cross-sectioning and path planning.

Motion and Path Planning for Additive Manufacturing
https://doi.org/10.1016/B978-0-443-15286-3.00010-2

Changing the slicing paradigm

So far, this book has presented the additive manufacturing process as one of sequential, monolithic steps that must be addressed to successfully construct an object: (1) an object is designed in CAD software and exported in a friendly representation like an STL, (2) this STL is then loaded into a slicing software package containing the algorithms presented in previous chapters and converted into the necessary g-code commands, and (3) these commands are finally taken to a 3D printer that constructs an object. This paradigm has existed alongside additive manufacturing since its inception and is a natural way to structure the process. However, this paradigm does have shortcomings, some of which were touched on in *Chapter 13: Pathing considerations for DED: Arc welding and laser welding*, sensor feedback control is vital to ensure high-quality builds, but in the current paradigm, the slicing software is not involved in the processing or utilization of the feedback, ultimately limiting the effectiveness of sensor feedback.

The argument for sensor feedback control is not limited to that situation. Large- and small-format systems—including polymer, wire filament, metal powder and more—can benefit from the use of sensor feedback. A review by Everton et al. showcases numerous examples of sensors used for feedback in metal AM, including image, acoustic, and force-based approaches [1]. The approaches for polymer are just as diverse. In general, these in-situ approaches attempt to identify and mitigate defects during the construction process. Naturally, a great deal of research has focused on the use of feedback in next-generation AM, but a fundamental problem remains. In these approaches, the feedback loop is not closed because it does not incorporate slicers. Typically, a separate software package handles the feedback analysis and construction modifications in tandem with the control system. However, why would the slicer that produced the pathing solution not be involved in the feedback loop? Would it not make more sense to have the software package that produced the initial pathing also be responsible for appropriately adjusting it, based on sensor feedback?

These fundamental questions have led to the development of a new paradigm, referred to as "real-time slicing" or "slice on demand" [2]. In this paradigm, rather than treat slicing as a sequential step in a process, slicing becomes an active component of the construction process. The slicer produces pathing for a single layer and sends this layer to the machine for construction. Once the layer has been constructed, sensor feedback is gathered, and the slicer ingests this feedback to adjust subsequent pathing. This process, as illustrated in Fig. 19.1, can be repeated throughout the construction so that the sensor feedback loop can be closed with respect to the slicing software. Closing the loop in this way allows the system to rely on the slicing package to produce necessary pathing adjustments, rather than develop an entirely new software package. The existing

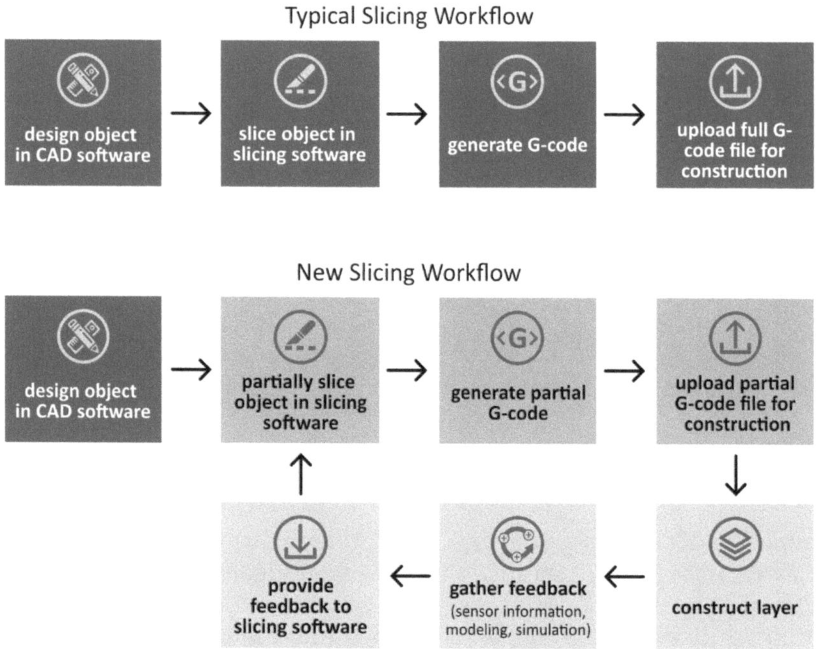

FIG. 19.1 Typical versus proposed slicing workflow. Originally, slicing is a sequential set of steps carried out in isolation. The proposed workflow seeks to integrate slicing with the construction process to support other software computation such as sensor feedback.

algorithms and processes discussed in the previous chapters can then be leveraged to produce a more advanced form of slicing.

To use this new paradigm, support from the control system of the 3D printer is necessary. This is a large obstacle because most machines are not configured to support the printing process in this way. However, the benefits are significant. For example, as mentioned in *Chapter 13: Pathing considerations for DED: Arc welding and laser welding*, sensor control is used in large-format DED systems to enable height control, which is essential for successful use of DED. These systems utilize a primitive form of feedback, a laser profilometer, and support from both the slicer and control system to accomplish the necessary behavior. Imagine what other types of quality control could be realized if such sensor feedback were incorporated in more advanced solutions, for more types of AM processes.

Sensor feedback

Laser profilometry can be leveraged for polymer systems in much the same way that it is used in metal-based systems [3]. However, in polymer systems, general height control is not necessary, due to the material

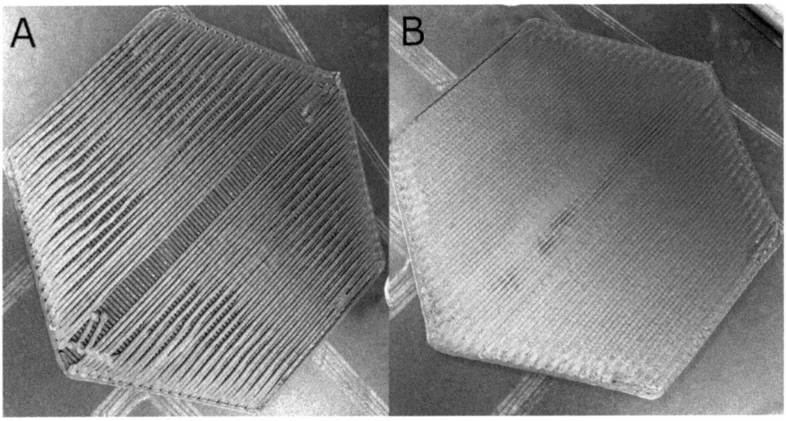

FIG. 19.2 Example of purposely inserted defects and process recovery; (A) clear examples of underfill; (B) recovery of the build a few layers later, using sensor feedback.

properties of polymers. That said, laser profilometry and height map generation can be used for the identification and mitigation of defects, such as underfill and overfill. An example of this is shown in Fig. 19.2. In this example, defects were purposely injected into the build and the system was allowed to subsequently identify and mitigate those defects.

Similar to the setup in DED, to achieve this utilization of feedback, it is not sufficient for the slicer to simply produce a g-code file. A slicer implementing slice-on-demand capabilities will also need to produce companion files. In this case, the companion files will describe the ideal construction of the object to use for comparisons to the generated height map.

Another straightforward application of sensor feedback is the use of thermal cameras. A slicer produces an estimate of the time it will take to print every layer so that the user can make speed adjustments before printing to ensure a desired layer time is achieved. However, this pre-planning of speeds and estimated layer times does not always reflect the real world, which can cause problems if the temperature of a layer goes outside of the acceptable range because of printing too quickly or too slowly. A more direct solution to the problem of layers times is the use of a thermal camera to collect data in real-time about how the object is cooling during the print. Using data from the thermal camera, the print speed can be increased to keep the layers warm enough for sufficient layer to layer bonding, or the speed can be decreased to ensure the object doesn't overheat and become distorted. For example, the object shown on the left side of Fig. 19.3 has been printed too quickly and collapses from excessive heat, but on the right side, the same object was successfully printed using thermal camera control to ensure the necessary layer times were achieved.

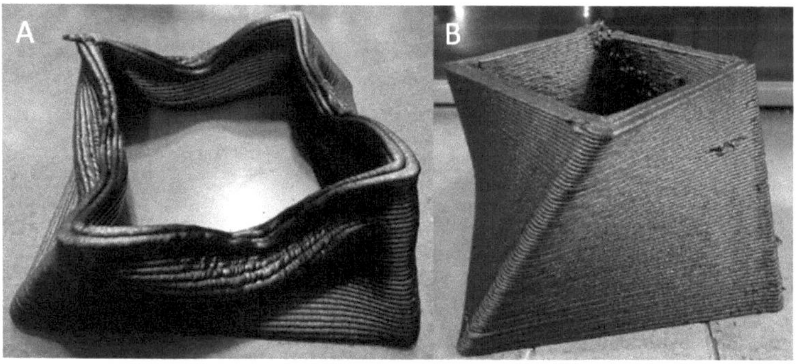

FIG. 19.3 A geometry printed with (right) and without (left) automatic layer time adjustment using a thermal camera; on the left, the print failed due to thermal buildup that prevented layers from properly cooling and hardening; on the right, the print succeeded because longer layer times allowed the thermal buildup to dissipate before continuing construction.

Support of experimental AM systems

In addition to better slicing via feedback, this paradigm also addresses issues that arise when using next-generation AM systems. For example, a rotary laser powder bed system (R-LPBF) is a next-gen powder bed system that constructs objects via a constant rotation of the bed. The bed itself is a torus, as shown in Fig. 19.4, with lasers offset at various locations around the build volume. From a hardware perspective, this system has several advantages over pre-existing powder bed solutions. One of the biggest advantages is a significantly increased throughput. A higher throughput means that much larger objects can be constructed in a reasonable amount of time. Additionally, those objects can be produced with the very high resolution that is common among powder bed systems. However, if a large object with a high resolution is being constructed, the amount of

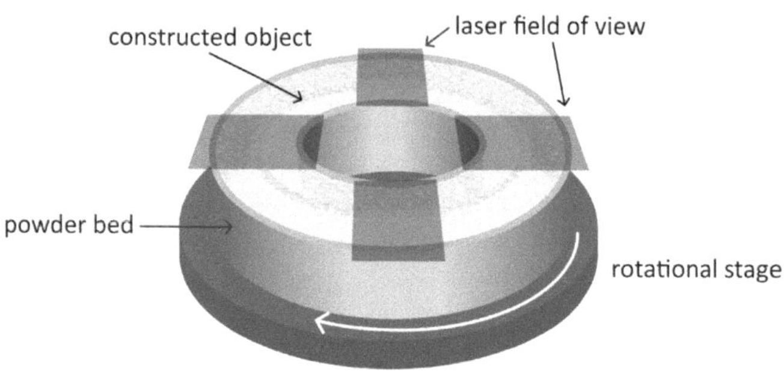

FIG. 19.4 Experimental rotary powder bed system. Such a system is not practical without the slice-on-demand paradigm, as computation times for path planning can be very large.

computation necessary to complete the slicing process is significantly increased. From a software perspective, this situation is quite complex.

In other, older, high-throughput systems, such as large-format polymer systems, there is usually a tradeoff between the size of the object and its resolution. As the object gets larger, the resolution reduces because the fundamental build process scales with size. This means that 3D printing a car on a large-format polymer system or miniature car on a desktop system are the same in terms of number of layers and total amount of pathing computed, and therefore slicing is essentially unaffected.

In the case of an R-LPBF, a typical layer height would be fractional millimeters and the total build volume can easily be a meter cubed. This easily amounts to 10,000+ layers with 10,000s of paths per layer to create a solid, dense object. Initial computations using commercially available software required 12–18 hours to produce the necessary pathing for such objects. For comparison, in the experimental slicing software the authors produce, the general rule is that slicing for a general object should take no more than a minute. Large or complex objects then should take no more than a few minutes.

While optimizations have been made that can reduce the time requirement for R-LPBF, the fact remains that there is an extraordinary amount of pathing to produce. While GPU optimization can significantly reduce the amount of time required—it has managed to reduce computation to only an hour or two—using an optimized GPU cannot completely compensate for the sheer amount of pathing computations. However, the slice-on-demand paradigm can be leveraged here. A single layer may only require a few seconds to produce the necessary pathing and even the fastest printer will require more time than this to complete a layer of construction. As a result, a slicer that implements the slice-on-demand paradigm can regularly stream g-code and hide the entire computation process behind the construction.

It is important to note that this new slicing paradigm assumes connectivity between the slicer, the control system, and the sensors of the machine. This connectivity, in turn, assumes network protocols are built into the slicing software. While the new paradigm is a significant departure from the norm, this connectivity alone is a relatively small shift that allows a natural evolution of the additive manufacturing process through software integration. Slicing should be one of several software packages working in tandem in a software ecosystem to effect the necessary control and feedback for an additive manufacturing system. This ecosystem is not and should not be limited to the slicer and control system. Ultimately, the slicing process is a purely geometric analysis, and while sensor feedback can assist this analysis, it does not explain the entire additive manufacturing process. With the necessary connectivity in place, new disciplines such as simulation, machine learning, and user interface design can be leveraged. Some of these possibilities are discussed in the following section.

Support and integration of simulations

One of the major areas of interest for integration is simulation. Simulation, such as finite element analysis (FEA), is regularly used in additive manufacturing to analyze the thermal or stress characteristics of the object that will be constructed [4]. FEA is also used extensively in topology optimization [5]. This is especially true in metal-based processes. However, simulation is usually accomplished separately from the slicing and build processes. The information learned is incorporated into the slicing in a roundabout way, via an informed user. This user manipulates the slicing parameters as necessary, because most slicers do not typically provide a way to ingest simulation information directly. However, by integrating a slicing package with simulation software, pathing solutions could be driven by finite element analysis.

For example, segments of pathing can be assigned material properties that are different from the surrounding pathing, based on the results of an FEA. To achieve this, the results of an FEA are reported in a grid structure, and this grid can be overlaid onto the pathing to automatically manipulate pathing properties. If an FEA of mechanical stress is conducted, the places in the pathing where additional stress is expected would be set to print with a denser material. This is illustrated by the simple square with an infill pattern shown in Fig. 19.5, where an FEA of stress in a similar object

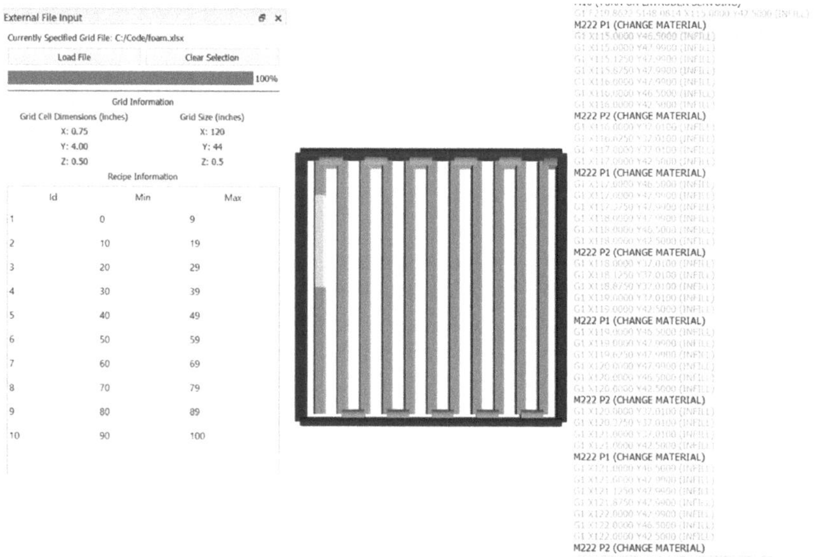

FIG. 19.5 Simple FEA example applied to pathing. In this case, FEA was conducted on stress. The FEA grid was overlaid onto the infill pathing and pathing modifications to density were driven by this grid.

has been imported and automatically applied to the infill pattern by the slicer. Despite the infill pattern being composed of solid lines, the segment highlighted in yellow shows that the pathing has been split. The highlighted segment will use a denser material than the surrounding pathing, because the FEA indicated that this area would exhibit higher stress. Although this example is simple, it expresses the potential for modifications that slicers could accomplish when integrated with simulations, machine learning models, or topological optimizations.

New visualization modalities

Novel visualization and interaction modalities are other examples of integrations that are made possible by the networking capability of an on-demand slicer. Many types of additive manufacturing and hybrid processes occlude the view of the operator. For example, in large-format direct energy deposition or wire-arc systems, the energy output is so great that protective enclosures must be utilized. With these enclosures, it is not possible to look directly at the construction point. Another example would be hybrid systems. When hybrid systems engage their subtractive process, the enclosure is typically flooded with coolant liquid that blocks the view of the object. In addition to these occlusion issues, general situational awareness is paramount, especially with large-format systems. If an unusual motion is commanded as part of the g-code with a large-format system, it's possible that the machine could be significantly damaged—an issue that is not generally present with desktop systems. One logical solution to improve awareness and provide context-appropriate information is the use of augmented reality (AR).

AR is the overlay of virtual information on the real world. There are two ways to accomplish this: wearables and handheld devices. Wearables are often specially made glasses or goggles that fit over the eyes and create a holographic projection within a specific field of view for the user. Conversely, handheld devices such as tablets or smartphones can accomplish similar capabilities by utilizing forward-facing cameras to render information on its screen while a user points it towards an area of interest. The most obvious difference between the two is whether the hands of the user are occupied or not, but for this discussion, the choice of device is not critical. AR devices provide a convenient way to visualize information that is contextually useful in 3D printing, such as models of g-code [4]. A simple example of what this might look like is shown in Fig. 19.6.

In this example, an AR device identifies a QR code that has been placed on a machine. This QR code has information that tells the AR device the location of the origin of the machine. Then, 3D models of the different layers of the constructed object and of the g-code commands are placed

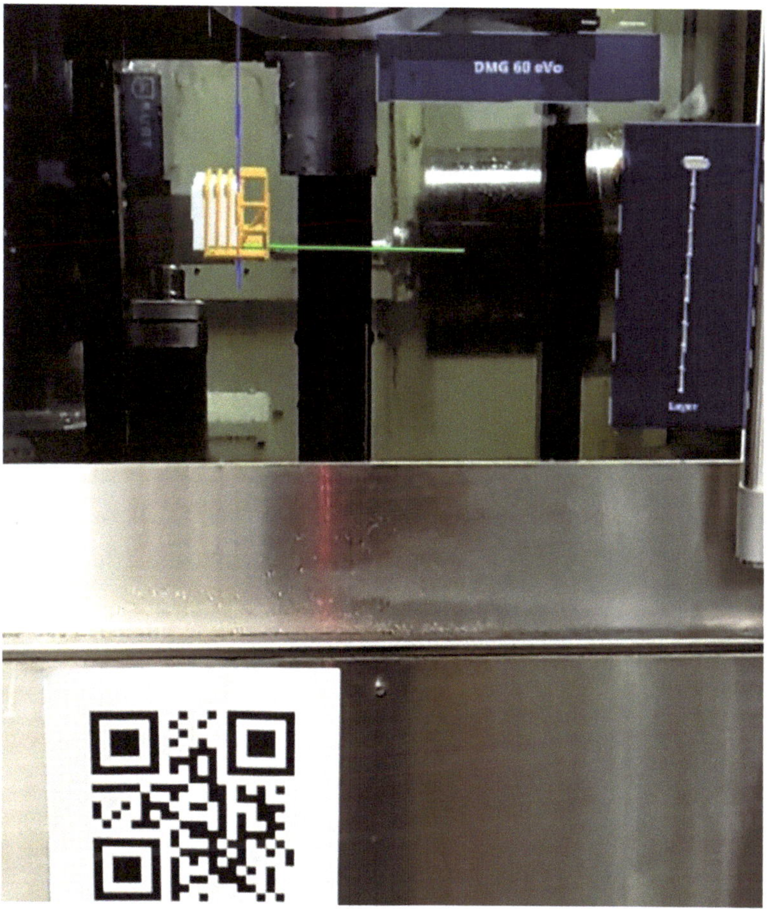

FIG. 19.6 Example augmented reality visualization. Many systems occlude the view of the object under construction. Augmented reality provides a way to visualize the g-code and the resulting changes to the object in a volumetrically appropriate way.

relative to the defined origin, so that they can be visualized in a spatially appropriate context. Importantly, these models are generated by the AR device and its associated software package, which process the g-code produced by a network-capable slicer. This is just one example of what is accomplishable when a slicer has network capabilities. When incorporated into a network, a slicer can become more than just a package for outputting pathing. It could also be used as a collaborative tool, connecting the user who is manipulating the slicer with the (potentially separate) operator of the machine.

The real challenge with using augmented reality and alternative visualization methodologies with slicing and 3D printing is the need to standardize visual information. Similar to how g-code standardizes the motions of a machine, visualized information needs to be standardized in its communication to support the widest number of machines possible. It should be noted that augmented reality's use with additive manufacturing is not limited to g-code visualization. Augmented reality has been used as a form of feedback and control, among other uses [6]. The point is that augmented reality's use with additive manufacturing is still very much an open question. While standardization and application of the technology is ongoing, the approach represents an active avenue of research.

Conclusion

This chapter has introduced cutting-edge concepts that may be applied to slicing. These concepts include a new slicing paradigm called "slicing on demand," as well as closed feedback loop control, support of next-generation systems, software ecosystem integration, and new visualization methodologies. Additionally, these concepts are topics of ongoing research and represent the evolution of the role of slicing in the AM workflow. As with all research topics, there is no telling whether any of these concepts will see widespread support. However, slicing software development cannot be ignored if the field of AM continues to grow. Each of these concepts exemplifies an expansion of the role of slicing and provides an avenue for future growth of AM as a serious industrial technology.

References

[1] S.K. Everton, M. Hirsch, P. Stravroulakis, R.K. Leach, A.T. Clare, Review of in-situ process monitoring and in-situ metrology for metal additive manufacturing, Mater. Des. 95 (2016) 431–445.

[2] M. Borish, A. Roschli, ORNL Slicer 2.0: towards a New Slicing Paradigm, in: In 2021 International Solid Freeform Fabrication Symposium, University of Texas at Austin, 2021.

[3] M. Borish, B.K. Post, A. Roschli, P.C. Chesser, L.J. Love, Real-time defect correction in large-scale polymer additive manufacturing via thermal imaging and laser profilometer, Proc. Manuf. 48 (2020) 625–633.

[4] Z. Luo, Y. Zhao, A survey of finite element analysis of temperature and thermal stress fields in powder bed fusion additive manufacturing, Addit. Manuf. 21 (2018) 318–332.

[5] J. Plocher, A. Panesar, Review on design and structural optimisation in additive manufacturing: towards next-generation lightweight structures, Mater. Des. 183 (2019) 108164.

[6] H. Lhachemi, A. Malik, R. Shorten, Augmented reality, cyber-physical systems, and feedback control for additive manufacturing: a review, IEEE Access 7 (2019) 50119–50135.

Further reading

M. Borish, J. Westfall, Additive and subtractive manufacturing augmented reality interface (ASMARI), in: In 2020 SoutheastCon, 2020, pp. 1–6.

APPENDIX

A

Geometry data file examples

An entire ASCII STL file of a 6 in (152.4 mm) cube generated with Solidworks.

```
solid 6in Cube ASCII
   facet normal -1.000000e+00 0.000000e+00 0.000000e+00
      outer loop
         vertex 0.000000e+00 0.000000e+00 1.524000e+02
         vertex 0.000000e+00 1.524000e+02 1.524000e+02
         vertex 0.000000e+00 0.000000e+00 0.000000e+00
      endloop
   endfacet
   facet normal -1.000000e+00 0.000000e+00 0.000000e+00
      outer loop
         vertex 0.000000e+00 0.000000e+00 0.000000e+00
         vertex 0.000000e+00 1.524000e+02 1.524000e+02
         vertex 0.000000e+00 1.524000e+02 0.000000e+00
      endloop
   endfacet
   facet normal -0.000000e+00 -1.000000e+00 -0.000000e+00
      outer loop
         vertex 1.524000e+02 0.000000e+00 1.524000e+02
         vertex 0.000000e+00 0.000000e+00 1.524000e+02
         vertex 1.524000e+02 0.000000e+00 0.000000e+00
      endloop
   endfacet
   facet normal 0.000000e+00 -1.000000e+00 0.000000e+00
      outer loop
         vertex 1.524000e+02 0.000000e+00 0.000000e+00
         vertex 0.000000e+00 0.000000e+00 1.524000e+02
         vertex 0.000000e+00 0.000000e+00 0.000000e+00
      endloop
```

```
endfacet
facet normal 1.000000e+00 0.000000e+00 0.000000e+00
   outer loop
      vertex 1.524000e+02 1.524000e+02 1.524000e+02
      vertex 1.524000e+02 0.000000e+00 1.524000e+02
      vertex 1.524000e+02 1.524000e+02 0.000000e+00
   endloop
endfacet
facet normal 1.000000e+00 0.000000e+00 0.000000e+00
   outer loop
      vertex 1.524000e+02 1.524000e+02 0.000000e+00
      vertex 1.524000e+02 0.000000e+00 1.524000e+02
      vertex 1.524000e+02 0.000000e+00 0.000000e+00
   endloop
endfacet
facet normal -0.000000e+00 1.000000e+00 0.000000e+00
   outer loop
      vertex 0.000000e+00 1.524000e+02 1.524000e+02
      vertex 1.524000e+02 1.524000e+02 1.524000e+02
      vertex 0.000000e+00 1.524000e+02 0.000000e+00
   endloop
endfacet
facet normal 0.000000e+00 1.000000e+00 0.000000e+00
   outer loop
      vertex 0.000000e+00 1.524000e+02 0.000000e+00
      vertex 1.524000e+02 1.524000e+02 1.524000e+02
      vertex 1.524000e+02 1.524000e+02 0.000000e+00
   endloop
endfacet
facet normal 0.000000e+00 -0.000000e+00 1.000000e+00
   outer loop
      vertex 1.524000e+02 0.000000e+00 1.524000e+02
      vertex 1.524000e+02 1.524000e+02 1.524000e+02
      vertex 0.000000e+00 0.000000e+00 1.524000e+02
   endloop
endfacet
facet normal 0.000000e+00 0.000000e+00 1.000000e+00
   outer loop
      vertex 0.000000e+00 0.000000e+00 1.524000e+02
      vertex 1.524000e+02 1.524000e+02 1.524000e+02
      vertex 0.000000e+00 1.524000e+02 1.524000e+02
   endloop
endfacet
facet normal -0.000000e+00 -0.000000e+00 -1.000000e+00
```

```
      outer loop
        vertex 1.524000e+02 1.524000e+02 0.000000e+00
        vertex 1.524000e+02 0.000000e+00 0.000000e+00
        vertex 0.000000e+00 1.524000e+02 0.000000e+00
      endloop
    endfacet
    facet normal 0.000000e+00 0.000000e+00 -1.000000e+00
      outer loop
        vertex 0.000000e+00 1.524000e+02 0.000000e+00
        vertex 1.524000e+02 0.000000e+00 0.000000e+00
        vertex 0.000000e+00 0.000000e+00 0.000000e+00
      endloop
    endfacet
endsolid
```

An entire binary STL file of a 6 in (152.4 mm) cube generated with Solidworks.

```
736f 6c69 6420 3669 6e20 4375 6265 2042
696e 6172 7920 2020 2020 2020 2020 2020
2020 2020 2020 2020 2020 2020 2020 2020
2020 2020 2020 2020 2020 2020 2020 2020
2020 2020 2020 2020 2020 2020 2020 2020
0c00 0000 0000 80bf 0000 0000 0000 0000
0000 0000 0000 0000 6766 1843 0000 0000
6766 1843 6766 1843 0000 0000 0000 0000
0000 0000 0000 0000 80bf 0000 0000 0000
0000 0000 0000 0000 0000 0000 0000 0000
0000 6766 1843 6766 1843 0000 0000 6766
1843 0000 0000 0000 0000 0080 0000 80bf
0000 0080 6766 1843 0000 0000 6766 1843
0000 0000 0000 0000 6766 1843 6766 1843
0000 0000 0000 0000 0000 0000 0000 0000
80bf 0000 0000 6766 1843 0000 0000 0000
0000 0000 0000 0000 0000 6766 1843 0000
0000 0000 0000 0000 0000 0000 0000 803f
0000 0000 0000 0000 6766 1843 6766 1843
6766 1843 6766 1843 0000 0000 6766 1843
6766 1843 6766 1843 0000 0000 0000 0000
803f 0000 0000 0000 0000 6766 1843 6766
1843 0000 0000 6766 1843 0000 0000 6766
1843 6766 1843 0000 0000 0000 0000 0000
0000 0080 0000 803f 0000 0000 0000 0000
6766 1843 6766 1843 6766 1843 6766 1843
6766 1843 0000 0000 6766 1843 0000 0000
```

```
0000 0000 0000 0000 803f 0000 0000 0000
0000 6766 1843 0000 0000 6766 1843 6766
1843 6766 1843 6766 1843 6766 1843 0000
0000 0000 0000 0000 0000 0080 0000 803f
6766 1843 0000 0000 6766 1843 6766 1843
6766 1843 6766 1843 0000 0000 0000 0000
6766 1843 0000 0000 0000 0000 0000 0000
803f 0000 0000 0000 0000 6766 1843 6766
1843 6766 1843 6766 1843 0000 0000 6766
1843 6766 1843 0000 0000 0080 0000 0080
0000 80bf 6766 1843 6766 1843 0000 0000
6766 1843 0000 0000 0000 0000 0000 0000
6766 1843 0000 0000 0000 0000 0000 0000
0000 0000 80bf 0000 0000 6766 1843 0000
0000 6766 1843 0000 0000 0000 0000 0000
0000 0000 0000 0000 0000 0000
```

An entire OBJ file of a 6 in (152.4 mm) cube generated with Autodesk Fusion 360.

```
# WaveFront *.obj file (generated by Autodesk ATF)
mtllib Body1.mtl
g Body1
v -76.200000 0.000000 -76.200000
v -76.200000 0.000000 76.200000
v -76.200000 152.400000 76.200000
v -76.200000 152.400000 -76.200000
v 76.200000 0.000000 -76.200000
v 76.200000 152.400000 -76.200000
v 76.200000 0.000000 76.200000
v 76.200000 152.400000 76.200000
vt 0.000000 0.000000 0.000000
vt 15.240000 0.000000 0.000000
vt 0.000000 15.240000 0.000000
vt 15.240000 15.240000 0.000000
vt -0.000000 0.000000 0.000000
vt 15.240000 0.000000 0.000000
vt 0.000000 15.240000 0.000000
vt 15.240000 15.240000 0.000000
vt 0.000000 0.000000 0.000000
vt 15.240000 0.000000 0.000000
vt 0.000000 15.240000 0.000000
vt 15.240000 15.240000 0.000000
vt 0.000000 0.000000 0.000000
vt 15.240000 0.000000 0.000000
```

```
vt 0.000000 15.240000 0.000000
vt 15.240000 15.240000 0.000000
vt 7.620000 7.620000 0.000000
vt -7.620000 7.620000 0.000000
vt 7.620000 -7.620000 0.000000
vt -7.620000 -7.620000 0.000000
vt -7.620000 -7.620000 0.000000
vt 7.620000 -7.620000 0.000000
vt -7.620000 7.6200a00 0.000000
vt 7.620000 7.620000 0.000000
vn -0.100000 0.000000 0.000000
vn -0.100000 0.000000 0.000000
vn -0.100000 0.000000 0.000000
vn -0.100000 0.000000 0.000000
vn 0.000000 0.000000 -0.100000
vn 0.000000 0.000000 -0.100000
vn 0.000000 0.000000 -0.100000
vn 0.000000 0.000000 -0.100000
vn 0.100000 0.000000 0.000000
vn 0.100000 0.000000 0.000000
vn 0.100000 0.000000 0.000000
vn 0.100000 0.000000 0.000000
vn 0.000000 -0.000000 0.100000
vn 0.000000 -0.000000 0.100000
vn 0.000000 -0.000000 0.100000
vn 0.000000 -0.000000 0.100000
vn 0.000000 0.100000 0.000000
vn 0.000000 0.100000 0.000000
vn 0.000000 0.100000 0.000000
vn 0.000000 0.100000 0.000000
vn 0.000000 -0.100000 0.000000
vn 0.000000 -0.100000 0.000000
vn 0.000000 -0.100000 0.000000
vn 0.000000 -0.100000 0.000000
usemtl Steel_-_Satin
f 1/1/1 2/2/2 4/3/3
f 4/3/3 2/2/2 3/4/4
f 5/5/5 1/6/6 6/7/7
f 6/7/7 1/6/6 4/8/8
f 7/9/9 5/10/10 8/11/11
f 8/11/11 5/10/10 6/12/12
f 2/13/13 7/14/14 3/15/15
f 3/15/15 7/14/14 8/16/16
f 8/17/17 6/18/18 3/19/19
```

```
f 3/19/19 6/18/18 4/20/20
f 7/21/21 2/22/22 5/23/23
f 5/23/23 2/22/22 1/24/24
# 8 vertices
# 24 texture params
# 24 normals
# 12 facets
# 1 groups
```

An entire AMF file of a 6 in (152.4 mm) cube generated with Solidworks.


```
<?xml version="1.0" encoding="UTF-8"?>
<amf unit="millimeter" version="1.1" xml:lang="en"><constellation
id="0"><instance objectid="276"/></constellation><object id="276"
type="model" materialid="1"><color><r>1</r><g>1</g><b>1</b>
</color><mesh><vertices><vertex><coordinates><x>-76.2</x><y>
-76.2</y><z>0</z></coordinates>
</vertex><vertex><coordinates><x>76.2</x><y>-76.2</y><z>0</z>
</coordinates></vertex><vertex><coordinates><x>-76.2</x>
<y>-76.2</y><z>152.4</z></coordinates>
</vertex><vertex><coordinates><x>76.2</x><y>-76.2</y>
<z>152.4</z></coordinates></vertex><vertex><coordinates>
<x>-76.2</x><y>76.2</y><z>0</z></coordinates>
</vertex><vertex><coordinates><x>76.2</x><y>76.2</y><z>0</z>
</coordinates></vertex><vertex><coordinates><x>-76.2</x>
<y>76.2</y><z>152.4</z></coordinates>
</vertex><vertex><coordinates><x>76.2</x><y>76.2</y><z>152.4</z>
</coordinates></vertex></vertices><volume><triangle>
<v1>2</v1><v2>6</v2><v3>0</v3></triangle><triangle>
<v1>6</v1><v2>4</v2><v3>0</v3></triangle><triangle>
<v1>1</v1><v2>2</v2><v3>0</v3></triangle><triangle>
<v1>4</v1><v2>1</v2><v3>0</v3></triangle><triangle>
<v1>3</v1><v2>2</v2><v3>1</v3></triangle><triangle>
<v1>5</v1><v2>3</v2><v3>1</v3></triangle><triangle>
<v1>4</v1><v2>5</v2><v3>1</v3></triangle><triangle>
<v1>3</v1><v2>7</v2><v3>2</v3></triangle><triangle>
<v1>7</v1><v2>6</v2><v3>2</v3></triangle><triangle>
<v1>5</v1><v2>7</v2><v3>3</v3></triangle><triangle>
<v1>6</v1><v2>7</v2><v3>4</v3></triangle><triangle>
<v1>7</v1><v2>5</v2><v3>4</v3></triangle></volume></mesh>
</object><material id="1"><metadata type="Name">6061
Alloy</metadata></material></amf>
```


An example of an entire 3MF file as provided by the 3MF Consortium.


```
<?xml version="1.0" encoding="UTF-8"?>
<model unit="millimeter" xml:lang="en-us"
xmlns:m="http://schemas.microsoft.com/3dmanufacturing/material/
2015/02"
xmlns="http://schemas.microsoft.com/3dmanufacturing/core/
2015/02"
```

```
  xmlns:vendor1="http://www.vendorwwebsite.com/3mf/
vendor13mfextension/2017/01">
  <metadata name="Copyright"> Microsoft Corporation 2013</metadata>
  <metadata name="Application">Microsoft 3D Builder</metadata>
  <metadata name="LicenseTerms">All rights reserved</metadata>
  <metadata name="Title">Cube</metadata>
  <metadata name="Designer">Microsoft Corporation</metadata>
  <metadata name="CreationDate">2013-10-07</metadata>
  <metadata name="ChangeHistory">2013-10-07 Initial model</
metadata>
  <metadata name="ModificationDate">2014-03-27</metadata>
  <metadata name="Description">Cube</metadata>
  <metadata name="vendor1:CustomMetadata1"
preserve="1">CE8A91FB-C44E-4F00-B634-
  BAA411465F6A</metadata>
  <resources>
  <basematerials id="1">
  <base name="Green" displaycolor="#21BB4CFF" />
  </basematerials>
  <object id="2" type="model" pid="1" pindex="0">
  <metadatagroup>
  <metadata name="vendor1:CustomMetadata2" preserve="true"
type="xs:string">03DAE6E4-24FF-4B20-
  97A1-7487AB9C1CB0</metadata>
  </metadatagroup>
  <mesh>
  <vertices>
  <vertex x="0" y="42.998" z="39.998" />
  <vertex x="39.998" y="42.998" z="39.998" />
  <vertex x="0" y="82.998" z="39.998" />
  <vertex x="39.998" y="82.998" z="0" />
  <vertex x="0" y="42.998" z="0" />
  <vertex x="0" y="82.998" z="0" />
  <vertex x="39.998" y="42.998" z="0" />
  <vertex x="39.998" y="82.998" z="39.998" />
  </vertices>
  <triangles>
  <triangle v1="0" v2="1" v3="2" />
  <triangle v1="3" v2="4" v3="5" />
  <triangle v1="4" v2="3" v3="6" />
  <triangle v1="7" v2="2" v3="1" />
  <triangle v1="4" v2="6" v3="1" />
  <triangle v1="4" v2="2" v3="5" />
  <triangle v1="7" v2="1" v3="6" />
```

```
<triangle v1="5" v2="2" v3="7" />
<triangle v1="4" v2="0" v3="2" />
<triangle v1="6" v2="3" v3="7" />
<triangle v1="1" v2="0" v3="4" />
<triangle v1="7" v2="3" v3="5" />
</triangles>
</mesh>
</object>
<object id="3" type="model">
<components>
<component objectid="2" />
</components>
</object>
</resources>
<build>
<item objectid="3" transform="1 0 0 0 1 0 0 0 1 -19.999
-62.998 0" >
<metadatagroup>
<metadata name="vendor1:CustomMetadata3"
type="xs:boolean">1</metadata>
</metadatagroup>
</item>
</build>
</model>
```

APPENDIX

B

G-code example files

Three layers of a cube g-code file for the RepRap syntax (.gcode). The first layer contains a solid skin, the second contains 25% dense infill, and the third has no infill.

```
;G-Code Syntax: RepRap
;Filename: Cube

;Slicing Parameters
;Nozzle Diameter: 0.4mm
;Layer Height: 0.2mm
;Default Extrusion Width: 0.4mm
;Perimeter Count: 1
;Inset Count: 2
;Upskin Count: 0
;Downskin Count: 1
;Skin Patern: Concentric

;Layer height: 0.2
M140 S60 ;SET BED TEMPERATURE
M190 S60 ;SET BED TEMPERATURE AND WAIT
M104 S210 T0 ;SET EXTRUDER 0 TEMPERATURE
M109 S210 T0 ;SET EXTRUDER 0 TEMPERATURE AND WAIT
M83 ;USE RELATIVE EXTRUSION DISTANCES
G92 E0 ;RESET FILAMENT AXIS TO 0


;LAYER COUNT:3
;BEGINNING LAYER: 1

;Nozzle 0
G1 F4200 E-3.2 ;RETRACT FILAMENT
```

```
G1 F3000 X0.0000 Y0.0000 Z0.2000 ;TRAVEL LIFT Z
G1 F6000 X73.4000 Y73.4000 Z0.4000 ;TRAVEL
G1 F3000 X73.4000 Y73.4000 Z0.2000 ;TRAVEL LOWER Z
G1 F0 E3.2 ;PRIME FILAMENT
G1 F3600 X126.6000 Y73.4000 E20.5078 ;SKIRT
G1 X126.6000 Y126.6000 E20.5078 ;SKIRT
G1 X73.4000 Y126.6000 E20.5078 ;SKIRT
G1 X73.4000 Y73.4000 E20.5078 ;SKIRT
G1 F4200 E-3.2 ;RETRACT FILAMENT
G1 F6000 X74.8000 Y74.8000 ;TRAVEL
G1 F0 E3.2 ;PRIME FILAMENT
G1 F3600 X125.2000 Y74.8000 E1.67631 ;PERIMETER
G1 X125.2000 Y125.2000 E1.67631 ;PERIMETER
G1 X74.8000 Y125.2000 E1.67631 ;PERIMETER
G1 X74.8000 Y74.8000 E1.67631 ;PERIMETER
G1 F6000 X75.2000 Y75.2000 ;TRAVEL
G1 F4200 X124.8000 Y75.2000 E1.6497 ;INSET
G1 X124.8000 Y124.8000 E1.6497 ;INSET
G1 X75.2000 Y124.8000 E1.6497 ;INSET
G1 X75.2000 Y75.2000 E1.6497 ;INSET
G1 F6000 X75.6000 Y75.6000 ;TRAVEL
G1 F4200 X124.4000 Y75.6000 E1.6231 ;INSET
G1 X124.4000 Y124.4000 E1.6231 ;INSET
G1 X75.6000 Y124.4000 E1.6231 ;INSET
G1 X75.6000 Y75.6000 E1.6231 ;INSET
G1 F6000 X76.0000 Y76.0000 ;TRAVEL
G1 F5400 X76.0000 Y124.0000 E1.59649 ;INFILL
G1 F4200 E-3.2 ;RETRACT FILAMENT
G1 F6000 X77.6000 Y124.0000 ;TRAVEL
G1 F0 E3.2 ;PRIME FILAMENT
G1 F5400 X77.6000 Y76.0000 E1.59649 ;INFILL
G1 F4200 E-3.2 ;RETRACT FILAMENT
G1 F6000 X79.2000 Y76.0000 ;TRAVEL
G1 F0 E3.2 ;PRIME FILAMENT
G1 F5400 X79.2000 Y124.0000 E1.59649 ;INFILL
G1 F4200 E-3.2 ;RETRACT FILAMENT
G1 F6000 X80.8000 Y124.0000 ;TRAVEL
G1 F0 E3.2 ;PRIME FILAMENT
G1 F5400 X80.8000 Y76.0000 E1.59649 ;INFILL
G1 F4200 E-3.2 ;RETRACT FILAMENT
G1 F6000 X82.4000 Y76.0000 ;TRAVEL
G1 F0 E3.2 ;PRIME FILAMENT
G1 F5400 X82.4000 Y124.0000 E1.59649 ;INFILL
G1 F4200 E-3.2 ;RETRACT FILAMENT
```

```
G1 F6000 X84.0000 Y124.0000 ;TRAVEL
G1 F0 E3.2 ;PRIME FILAMENT
G1 F5400 X84.0000 Y76.0000 E1.59649 ;INFILL
G1 F4200 E-3.2 ;RETRACT FILAMENT
G1 F6000 X85.6000 Y76.0000 ;TRAVEL
G1 F0 E3.2 ;PRIME FILAMENT
G1 F5400 X85.6000 Y124.0000 E1.59649 ;INFILL
G1 F4200 E-3.2 ;RETRACT FILAMENT
G1 F6000 X87.2000 Y124.0000 ;TRAVEL
G1 F0 E3.2 ;PRIME FILAMENT
G1 F5400 X87.2000 Y76.0000 E1.59649 ;INFILL
G1 F4200 E-3.2 ;RETRACT FILAMENT
G1 F6000 X88.8000 Y76.0000 ;TRAVEL
G1 F0 E3.2 ;PRIME FILAMENT
G1 F5400 X88.8000 Y124.0000 E1.59649 ;INFILL
G1 F4200 E-3.2 ;RETRACT FILAMENT
G1 F6000 X90.4000 Y124.0000 ;TRAVEL
G1 F0 E3.2 ;PRIME FILAMENT
G1 F5400 X90.4000 Y76.0000 E1.59649 ;INFILL
G1 F4200 E-3.2 ;RETRACT FILAMENT
G1 F6000 X92.0000 Y76.0000 ;TRAVEL
G1 F0 E3.2 ;PRIME FILAMENT
G1 F5400 X92.0000 Y124.0000 E1.59649 ;INFILL
G1 F4200 E-3.2 ;RETRACT FILAMENT
G1 F6000 X93.6000 Y124.0000 ;TRAVEL
G1 F0 E3.2 ;PRIME FILAMENT
G1 F5400 X93.6000 Y76.0000 E1.59649 ;INFILL
G1 F4200 E-3.2 ;RETRACT FILAMENT
G1 F6000 X95.2000 Y76.0000 ;TRAVEL
G1 F0 E3.2 ;PRIME FILAMENT
G1 F5400 X95.2000 Y124.0000 E1.59649 ;INFILL
G1 F4200 E-3.2 ;RETRACT FILAMENT
G1 F6000 X96.8000 Y124.0000 ;TRAVEL
G1 F0 E3.2 ;PRIME FILAMENT
G1 F5400 X96.8000 Y76.0000 E1.59649 ;INFILL
G1 F4200 E-3.2 ;RETRACT FILAMENT
G1 F6000 X98.4000 Y76.0000 ;TRAVEL
G1 F0 E3.2 ;PRIME FILAMENT
G1 F5400 X98.4000 Y124.0000 E1.59649 ;INFILL
G1 F4200 E-3.2 ;RETRACT FILAMENT
G1 F6000 X100.0000 Y124.0000 ;TRAVEL
G1 F0 E3.2 ;PRIME FILAMENT
G1 F5400 X100.0000 Y76.0000 E1.59649 ;INFILL
G1 F4200 E-3.2 ;RETRACT FILAMENT
```

```
G1 F6000 X101.6000 Y76.0000 ;TRAVEL
G1 F0 E3.2 ;PRIME FILAMENT
G1 F5400 X101.6000 Y124.0000 E1.59649 ;INFILL
G1 F4200 E-3.2 ;RETRACT FILAMENT
G1 F6000 X103.2000 Y124.0000 ;TRAVEL
G1 F0 E3.2 ;PRIME FILAMENT
G1 F5400 X103.2000 Y76.0000 E1.59649 ;INFILL
G1 F4200 E-3.2 ;RETRACT FILAMENT
G1 F6000 X104.8000 Y76.0000 ;TRAVEL
G1 F0 E3.2 ;PRIME FILAMENT
G1 F5400 X104.8000 Y124.0000 E1.59649 ;INFILL
G1 F4200 E-3.2 ;RETRACT FILAMENT
G1 F6000 X106.4000 Y124.0000 ;TRAVEL
G1 F0 E3.2 ;PRIME FILAMENT
G1 F5400 X106.4000 Y76.0000 E1.59649 ;INFILL
G1 F4200 E-3.2 ;RETRACT FILAMENT
G1 F6000 X108.0000 Y76.0000 ;TRAVEL
G1 F0 E3.2 ;PRIME FILAMENT
G1 F5400 X108.0000 Y124.0000 E1.59649 ;INFILL
G1 F4200 E-3.2 ;RETRACT FILAMENT
G1 F6000 X109.6000 Y124.0000 ;TRAVEL
G1 F0 E3.2 ;PRIME FILAMENT
G1 F5400 X109.6000 Y76.0000 E1.59649 ;INFILL
G1 F4200 E-3.2 ;RETRACT FILAMENT
G1 F6000 X111.2000 Y76.0000 ;TRAVEL
G1 F0 E3.2 ;PRIME FILAMENT
G1 F5400 X111.2000 Y124.0000 E1.59649 ;INFILL
G1 F4200 E-3.2 ;RETRACT FILAMENT
G1 F6000 X112.8000 Y124.0000 ;TRAVEL
G1 F0 E3.2 ;PRIME FILAMENT
G1 F5400 X112.8000 Y76.0000 E1.59649 ;INFILL
G1 F4200 E-3.2 ;RETRACT FILAMENT
G1 F6000 X114.4000 Y76.0000 ;TRAVEL
G1 F0 E3.2 ;PRIME FILAMENT
G1 F5400 X114.4000 Y124.0000 E1.59649 ;INFILL
G1 F4200 E-3.2 ;RETRACT FILAMENT
G1 F6000 X116.0000 Y124.0000 ;TRAVEL
G1 F0 E3.2 ;PRIME FILAMENT
G1 F5400 X116.0000 Y76.0000 E1.59649 ;INFILL
G1 F4200 E-3.2 ;RETRACT FILAMENT
G1 F6000 X117.6000 Y76.0000 ;TRAVEL
G1 F0 E3.2 ;PRIME FILAMENT
G1 F5400 X117.6000 Y124.0000 E1.59649 ;INFILL
G1 F4200 E-3.2 ;RETRACT FILAMENT
```

```
G1 F6000 X119.2000 Y124.0000 ;TRAVEL
G1 F0 E3.2 ;PRIME FILAMENT
G1 F5400 X119.2000 Y76.0000 E1.59649 ;INFILL
G1 F4200 E-3.2 ;RETRACT FILAMENT
G1 F6000 X120.8000 Y76.0000 ;TRAVEL
G1 F0 E3.2 ;PRIME FILAMENT
G1 F5400 X120.8000 Y124.0000 E1.59649 ;INFILL
G1 F4200 E-3.2 ;RETRACT FILAMENT
G1 F6000 X122.4000 Y124.0000 ;TRAVEL
G1 F0 E3.2 ;PRIME FILAMENT
G1 F5400 X122.4000 Y76.0000 E1.59649 ;INFILL

;BEGINNING LAYER: 2
G1 F4200 E-3.2 ;RETRACT FILAMENT

;Nozzle 0
G1 F6000 X125.2000 Y74.8000 Z0.4000 ;TRAVEL
G1 F0 E3.2 ;PRIME FILAMENT
G1 F3600 X125.2000 Y125.2000 E1.67631 ;PERIMETER
G1 X74.8000 Y125.2000 E1.67631 ;PERIMETER
G1 X74.8000 Y74.8000 E1.67631 ;PERIMETER
G1 X125.2000 Y74.8000 E1.67631 ;PERIMETER
G1 F6000 X124.8000 Y75.2000 ;TRAVEL
G1 F4200 X124.8000 Y124.8000 E1.6497 ;INSET
G1 X75.2000 Y124.8000 E1.6497 ;INSET
G1 X75.2000 Y75.2000 E1.6497 ;INSET
G1 X124.8000 Y75.2000 E1.6497 ;INSET
G1 F6000 X124.4000 Y75.6000 ;TRAVEL
G1 F4200 X124.4000 Y124.4000 E1.6231 ;INSET
G1 X75.6000 Y124.4000 E1.6231 ;INSET
G1 X75.6000 Y75.6000 E1.6231 ;INSET
G1 X124.4000 Y75.6000 E1.6231 ;INSET
G1 F6000 X124.0000 Y75.9994 ;TRAVEL
G1 F5400 X75.9997 Y75.9993 E1.59649 ;INFILL
G1 F4200 E-3.2 ;RETRACT FILAMENT
G1 F6000 X75.9997 Y77.5993 ;TRAVEL
G1 F0 E3.2 ;PRIME FILAMENT
G1 F5400 X124.0000 Y77.5994 E1.59649 ;INFILL
G1 F4200 E-3.2 ;RETRACT FILAMENT
G1 F6000 X124.0000 Y79.1994 ;TRAVEL
G1 F0 E3.2 ;PRIME FILAMENT
G1 F5400 X75.9997 Y79.1993 E1.59649 ;INFILL
G1 F4200 E-3.2 ;RETRACT FILAMENT
G1 F6000 X75.9997 Y80.7993 ;TRAVEL
```

```
G1 F0 E3.2 ;PRIME FILAMENT
G1 F5400 X124.0000 Y80.7994 E1.59649 ;INFILL
G1 F4200 E-3.2 ;RETRACT FILAMENT
G1 F6000 X124.0000 Y82.3994 ;TRAVEL
G1 F0 E3.2 ;PRIME FILAMENT
G1 F5400 X75.9997 Y82.3993 E1.59649 ;INFILL
G1 F4200 E-3.2 ;RETRACT FILAMENT
G1 F6000 X75.9997 Y83.9993 ;TRAVEL
G1 F0 E3.2 ;PRIME FILAMENT
G1 F5400 X124.0000 Y83.9994 E1.59649 ;INFILL
G1 F4200 E-3.2 ;RETRACT FILAMENT
G1 F6000 X124.0000 Y85.5994 ;TRAVEL
G1 F0 E3.2 ;PRIME FILAMENT
G1 F5400 X75.9997 Y85.5993 E1.59649 ;INFILL
G1 F4200 E-3.2 ;RETRACT FILAMENT
G1 F6000 X75.9997 Y87.1993 ;TRAVEL
G1 F0 E3.2 ;PRIME FILAMENT
G1 F5400 X124.0000 Y87.1994 E1.59649 ;INFILL
G1 F4200 E-3.2 ;RETRACT FILAMENT
G1 F6000 X124.0000 Y88.7994 ;TRAVEL
G1 F0 E3.2 ;PRIME FILAMENT
G1 F5400 X75.9997 Y88.7993 E1.59649 ;INFILL
G1 F4200 E-3.2 ;RETRACT FILAMENT
G1 F6000 X75.9997 Y90.3993 ;TRAVEL
G1 F0 E3.2 ;PRIME FILAMENT
G1 F5400 X124.0000 Y90.3994 E1.59649 ;INFILL
G1 F4200 E-3.2 ;RETRACT FILAMENT
G1 F6000 X124.0000 Y91.9994 ;TRAVEL
G1 F0 E3.2 ;PRIME FILAMENT
G1 F5400 X75.9997 Y91.9993 E1.59649 ;INFILL
G1 F4200 E-3.2 ;RETRACT FILAMENT
G1 F6000 X75.9997 Y93.5993 ;TRAVEL
G1 F0 E3.2 ;PRIME FILAMENT
G1 F5400 X124.0000 Y93.5994 E1.59649 ;INFILL
G1 F4200 E-3.2 ;RETRACT FILAMENT
G1 F6000 X124.0000 Y95.1994 ;TRAVEL
G1 F0 E3.2 ;PRIME FILAMENT
G1 F5400 X75.9997 Y95.1993 E1.59649 ;INFILL
G1 F4200 E-3.2 ;RETRACT FILAMENT
G1 F6000 X75.9997 Y96.7993 ;TRAVEL
G1 F0 E3.2 ;PRIME FILAMENT
G1 F5400 X124.0000 Y96.7994 E1.59649 ;INFILL
G1 F4200 E-3.2 ;RETRACT FILAMENT
G1 F6000 X124.0000 Y98.3994 ;TRAVEL
```

```
G1 F0 E3.2 ;PRIME FILAMENT
G1 F5400 X75.9997 Y98.3993 E1.59649 ;INFILL
G1 F4200 E-3.2 ;RETRACT FILAMENT
G1 F6000 X75.9996 Y99.9993 ;TRAVEL
G1 F0 E3.2 ;PRIME FILAMENT
G1 F5400 X124.0000 Y99.9994 E1.59649 ;INFILL
G1 F4200 E-3.2 ;RETRACT FILAMENT
G1 F6000 X124.0000 Y101.5990 ;TRAVEL
G1 F0 E3.2 ;PRIME FILAMENT
G1 F5400 X75.9996 Y101.5990 E1.59649 ;INFILL
G1 F4200 E-3.2 ;RETRACT FILAMENT
G1 F6000 X75.9996 Y103.1990 ;TRAVEL
G1 F0 E3.2 ;PRIME FILAMENT
G1 F5400 X124.0000 Y103.1990 E1.59649 ;INFILL
G1 F4200 E-3.2 ;RETRACT FILAMENT
G1 F6000 X124.0000 Y104.7990 ;TRAVEL
G1 F0 E3.2 ;PRIME FILAMENT
G1 F5400 X75.9996 Y104.7990 E1.59649 ;INFILL
G1 F4200 E-3.2 ;RETRACT FILAMENT
G1 F6000 X75.9996 Y106.3990 ;TRAVEL
G1 F0 E3.2 ;PRIME FILAMENT
G1 F5400 X124.0000 Y106.3990 E1.59649 ;INFILL
G1 F4200 E-3.2 ;RETRACT FILAMENT
G1 F6000 X124.0000 Y107.9990 ;TRAVEL
G1 F0 E3.2 ;PRIME FILAMENT
G1 F5400 X75.9996 Y107.9990 E1.59649 ;INFILL
G1 F4200 E-3.2 ;RETRACT FILAMENT
G1 F6000 X75.9996 Y109.5990 ;TRAVEL
G1 F0 E3.2 ;PRIME FILAMENT
G1 F5400 X124.0000 Y109.5990 E1.59649 ;INFILL
G1 F4200 E-3.2 ;RETRACT FILAMENT
G1 F6000 X124.0000 Y111.1990 ;TRAVEL
G1 F0 E3.2 ;PRIME FILAMENT
G1 F5400 X75.9996 Y111.1990 E1.59649 ;INFILL
G1 F4200 E-3.2 ;RETRACT FILAMENT
G1 F6000 X75.9996 Y112.7990 ;TRAVEL
G1 F0 E3.2 ;PRIME FILAMENT
G1 F5400 X124.0000 Y112.7990 E1.59649 ;INFILL
G1 F4200 E-3.2 ;RETRACT FILAMENT
G1 F6000 X124.0000 Y114.3990 ;TRAVEL
G1 F0 E3.2 ;PRIME FILAMENT
G1 F5400 X75.9996 Y114.3990 E1.59649 ;INFILL
G1 F4200 E-3.2 ;RETRACT FILAMENT
G1 F6000 X75.9996 Y115.9990 ;TRAVEL
```

```
G1 F0 E3.2 ;PRIME FILAMENT
G1 F5400 X124.0000 Y115.9990 E1.59649 ;INFILL
G1 F4200 E-3.2 ;RETRACT FILAMENT
G1 F6000 X124.0000 Y117.5990 ;TRAVEL
G1 F0 E3.2 ;PRIME FILAMENT
G1 F5400 X75.9996 Y117.5990 E1.59649 ;INFILL
G1 F4200 E-3.2 ;RETRACT FILAMENT
G1 F6000 X75.9996 Y119.1990 ;TRAVEL
G1 F0 E3.2 ;PRIME FILAMENT
G1 F5400 X124.0000 Y119.1990 E1.59649 ;INFILL
G1 F4200 E-3.2 ;RETRACT FILAMENT
G1 F6000 X124.0000 Y120.7990 ;TRAVEL
G1 F0 E3.2 ;PRIME FILAMENT
G1 F5400 X75.9996 Y120.7990 E1.59649 ;INFILL
G1 F4200 E-3.2 ;RETRACT FILAMENT
G1 F6000 X75.9996 Y122.3990 ;TRAVEL
G1 F0 E3.2 ;PRIME FILAMENT
G1 F5400 X124.0000 Y122.3990 E1.59649 ;INFILL

;BEGINNING LAYER: 3
G1 F4200 E-3.2 ;RETRACT FILAMENT

;Nozzle 0
G1 F6000 X74.8000 Y74.8000 Z0.6000 ;TRAVEL
G1 F0 E3.2 ;PRIME FILAMENT
G1 F3600 X125.2000 Y74.8000 E1.67631 ;PERIMETER
G1 X125.2000 Y125.2000 E1.67631 ;PERIMETER
G1 X74.8000 Y125.2000 E1.67631 ;PERIMETER
G1 X74.8000 Y74.8000 E1.67631 ;PERIMETER
G1 F6000 X75.2000 Y75.2000 ;TRAVEL
G1 F4200 X124.8000 Y75.2000 E1.6497 ;INSET
G1 X124.8000 Y124.8000 E1.6497 ;INSET
G1 X75.2000 Y124.8000 E1.6497 ;INSET
G1 X75.2000 Y75.2000 E1.6497 ;INSET
G1 F6000 X75.6000 Y75.6000 ;TRAVEL
G1 F4200 X124.4000 Y75.6000 E1.6231 ;INSET
G1 X124.4000 Y124.4000 E1.6231 ;INSET
G1 X75.6000 Y124.4000 E1.6231 ;INSET
G1 X75.6000 Y75.6000 E1.6231 ;INSET

M104 S0 ;TURN EXTRUDER OFF
M140 S0 ;TURN BED OFF
M84 ;DISABLE MOTORS
M106 S0 ;DISABLE COOLING FAN
```

Three layers of a cube g-code file for the Cincinnati BAAM syntax (.nc). The first layer contains a solid skin, the second contains 25% dense infill, and the third has no infill.

```
(G-Code Syntax: Cincinnati BAAM)
(Filename: Cube)

(SAFETY BLOCK - ESTABLISH OPERATIONAL MODES)
M11 (EXIT SERVOING MODE)
M66 L0.177 (SET ACCELERATION)
M16 (DEFAULT SPINDLE ADJUSTMENT)
M270 L1 (SET FEEDRATE MULTIPLIER TO 100%)
G1 F120 (SET INITIAL FEEDRATE)
M0 (WAIT FOR USER)
G4 P0.2500 (DWELL)
G0 Z0 (LIFT Z FOR SAFETY)
G0 W0.0000 (SET INITIAL W HEIGHT)
M69 F200 P45 S3 (PURGE)
M0 (WAIT FOR USER)

(LAYER COUNT: 3)
(BEGINNING LAYER: 1)
G0 W-0.1500 (MOVE W - LAYER CHANGE)
M1 (OPTIONAL STOP - LAYER CHANGE)

(Nozzle 0)
M12 (PERIMETER SPINDLE ADJUSTMENT)
M66 L0.2655 (SET ACCELERATION)
G0 X114.1700 Y36.6700 Z0.0000 (TRAVEL)
G0 X114.1700 Y36.6700 Z-15.2500 (TRAVEL LOWER Z)
M64 L0.7 (TURN TAMPER ON)
M11 (TURN OFF EXTRUDER SERVOING)
M3 S300 (TURN EXTRUDER ON)
G4 P0.0500 (DWELL)
M10 (TURN ON EXTRUDER SERVOING)
G1 F452.6716 S274.3464 X125.8300 Y36.6700 (PERIMETER)
G1 X125.8300 Y48.3300 (PERIMETER)
G1 X114.1700 Y48.3300 (PERIMETER)
G1 X114.1700 Y36.6700 (PERIMETER)
G1 F740.7354 S0.0000 X117.1700 Y36.6700 (PERIMETER FORWARD TIP
WIPE)
M65 (TURN TAMPER OFF)
M5 (TURN EXTRUDER OFF)
M13 (INSET SPINDLE ADJUSTMENT)
M66 L0.1991 (SET ACCELERATION)
```

```
G0 X117.1700 Y36.6700 Z-14.7500 (TRAVEL LIFT Z)
G0 X114.8500 Y37.3500 (TRAVEL)
G0 X114.8500 Y37.3500 Z-15.2500 (TRAVEL LOWER Z)
M64 L0.7 (TURN TAMPER ON)
M11 (TURN OFF EXTRUDER SERVOING)
M3 S300 (TURN EXTRUDER ON)
G4 P0.0500 (DWELL)
M10 (TURN ON EXTRUDER SERVOING)
G1 F452.6716 S274.3464 X125.1500 Y37.3500 (INSET)
G1 X125.1500 Y47.6500 (INSET)
G1 X114.8500 Y47.6500 (INSET)
G1 X114.8500 Y37.3500 (INSET)
G1 F740.7354 S0.0000 X117.8500 Y37.3500 (INSET FORWARD TIP WIPE)
M65 (TURN TAMPER OFF)
M5 (TURN EXTRUDER OFF)
M66 L0.1991 (SET ACCELERATION)
G0 X117.8500 Y37.3500 Z-14.7500 (TRAVEL LIFT Z)
G0 X114.5100 Y37.0100 (TRAVEL)
G0 X114.5100 Y37.0100 Z-15.2500 (TRAVEL LOWER Z)
M64 L0.7 (TURN TAMPER ON)
M11 (TURN OFF EXTRUDER SERVOING)
M3 S300 (TURN EXTRUDER ON)
G4 P0.0500 (DWELL)
M10 (TURN ON EXTRUDER SERVOING)
G1 F452.6716 S274.3464 X125.4900 Y37.0100 (INSET)
G1 X125.4900 Y47.9900 (INSET)
G1 X114.5100 Y47.9900 (INSET)
G1 X114.5100 Y37.0100 (INSET)
G1 F740.7354 S0.0000 X117.5100 Y37.0100 (INSET FORWARD TIP WIPE)
M65 (TURN TAMPER OFF)
M5 (TURN EXTRUDER OFF)
M15 (SKIN SPINDLE ADJUSTMENT)
M66 L0.177 (SET ACCELERATION)
G0 X117.5100 Y37.0100 Z-14.7500 (TRAVEL LIFT Z)
G0 X117.6200 Y37.6900 (TRAVEL)
G0 X117.6200 Y37.6900 Z-15.2500 (TRAVEL LOWER Z)
M64 L0.7 (TURN TAMPER ON)
M11 (TURN OFF EXTRUDER SERVOING)
M3 S300 (TURN EXTRUDER ON)
G4 P0.0500 (DWELL)
M10 (TURN ON EXTRUDER SERVOING)
G1 F452.6716 S240.0531 X117.6200 Y47.3100 (SKIN)
G1 X117.2800 Y47.3100 (SKIN)
G1 X117.2800 Y37.6900 (SKIN)
```

```
G1 X116.9400 Y37.6900 (SKIN)
G1 X116.9400 Y47.3100 (SKIN)
G1 X116.6000 Y47.3100 (SKIN)
G1 X116.6000 Y37.6900 (SKIN)
G1 X116.2600 Y37.6900 (SKIN)
G1 X116.2600 Y47.3100 (SKIN)
G1 X115.9200 Y47.3100 (SKIN)
G1 X115.9200 Y37.6900 (SKIN)
G1 X115.5800 Y37.6900 (SKIN)
G1 X115.5800 Y47.3100 (SKIN)
G1 X115.2400 Y47.3100 (SKIN)
G1 X115.2400 Y37.6900 (SKIN)
G1 F740.7354 S0.0000 X115.2400 Y37.3500 (SKIN FORWARD TIP WIPE)
G1 X114.8500 Y37.3500 (SKIN FORWARD TIP WIPE)
G1 X114.8500 Y39.6200 (SKIN FORWARD TIP WIPE)
M65 (TURN TAMPER OFF)
M5 (TURN EXTRUDER OFF)
M66 L0.177 (SET ACCELERATION)
G0 X114.8500 Y39.6200 Z-14.7500 (TRAVEL LIFT Z)
G0 X117.9600 Y37.6900 (TRAVEL)
G0 X117.9600 Y37.6900 Z-15.2500 (TRAVEL LOWER Z)
M64 L0.7 (TURN TAMPER ON)
M11 (TURN OFF EXTRUDER SERVOING)
M3 S300 (TURN EXTRUDER ON)
G4 P0.0500 (DWELL)
M10 (TURN ON EXTRUDER SERVOING)
G1 F452.6716 S240.0531 X117.9600 Y47.3100 (SKIN)
G1 X118.3000 Y47.3100 (SKIN)
G1 X118.3000 Y37.6900 (SKIN)
G1 X118.6400 Y37.6900 (SKIN)
G1 X118.6400 Y47.3100 (SKIN)
G1 X118.9800 Y47.3100 (SKIN)
G1 X118.9800 Y37.6900 (SKIN)
G1 X119.3200 Y37.6900 (SKIN)
G1 X119.3200 Y47.3100 (SKIN)
G1 X119.6600 Y47.3100 (SKIN)
G1 X119.6600 Y37.6900 (SKIN)
G1 X120.0000 Y37.6900 (SKIN)
G1 X120.0000 Y47.3100 (SKIN)
G1 X120.3400 Y47.3100 (SKIN)
G1 X120.3400 Y37.6900 (SKIN)
G1 X120.6800 Y37.6900 (SKIN)
G1 X120.6800 Y47.3100 (SKIN)
G1 X121.0200 Y47.3100 (SKIN)
```

```
G1 X121.0200 Y37.6900 (SKIN)
G1 X121.3600 Y37.6900 (SKIN)
G1 X121.3600 Y47.3100 (SKIN)
G1 X121.7000 Y47.3100 (SKIN)
G1 X121.7000 Y37.6900 (SKIN)
G1 X122.0400 Y37.6900 (SKIN)
G1 X122.0400 Y47.3100 (SKIN)
G1 X122.3800 Y47.3100 (SKIN)
G1 X122.3800 Y37.6900 (SKIN)
G1 X122.7200 Y37.6900 (SKIN)
G1 X122.7200 Y47.3100 (SKIN)
G1 X123.0600 Y47.3100 (SKIN)
G1 X123.0600 Y37.6900 (SKIN)
G1 X123.4000 Y37.6900 (SKIN)
G1 X123.4000 Y47.3100 (SKIN)
G1 X123.7400 Y47.3100 (SKIN)
G1 X123.7400 Y37.6900 (SKIN)
G1 X124.0800 Y37.6900 (SKIN)
G1 X124.0800 Y47.3100 (SKIN)
G1 X124.4200 Y47.3100 (SKIN)
G1 X124.4200 Y37.6900 (SKIN)
G1 X124.7600 Y37.6900 (SKIN)
G1 X124.7600 Y47.3100 (SKIN)
G1 F740.7354 S0.0000 X124.7600 Y47.6500 (SKIN FORWARD TIP WIPE)
G1 X125.1500 Y47.6500 (SKIN FORWARD TIP WIPE)
G1 X125.1500 Y45.3800 (SKIN FORWARD TIP WIPE)
M65 (TURN TAMPER OFF)
M5 (TURN EXTRUDER OFF)

(BEGINNING LAYER: 2)
G0 W-0.3000 (MOVE W - LAYER CHANGE)
M1 (OPTIONAL STOP - LAYER CHANGE)

(Nozzle 0)
M12 (PERIMETER SPINDLE ADJUSTMENT)
M66 L0.2655 (SET ACCELERATION)
G0 X125.1500 Y45.3800 Z-14.7500 (TRAVEL LIFT Z)
G0 X125.8300 Y48.3300 (TRAVEL)
G0 X125.8300 Y48.3300 Z-15.2500 (TRAVEL LOWER Z)
M64 L0.7 (TURN TAMPER ON)
M11 (TURN OFF EXTRUDER SERVOING)
M3 S300 (TURN EXTRUDER ON)
G4 P0.0500 (DWELL)
M10 (TURN ON EXTRUDER SERVOING)
G1 F182.6948 S110.7241 X114.1700 Y48.3300 (PERIMETER)
```

```
G1 X114.1700 Y36.6700 (PERIMETER)
G1 X125.8300 Y36.6700 (PERIMETER)
G1 X125.8300 Y48.3300 (PERIMETER)
G1 F298.9551 S0.0000 X122.8300 Y48.3300 (PERIMETER FORWARD TIP
WIPE)
M65 (TURN TAMPER OFF)
M5 (TURN EXTRUDER OFF)
M13 (INSET SPINDLE ADJUSTMENT)
M66 L0.1991 (SET ACCELERATION)
G0 X122.8300 Y48.3300 Z-14.7500 (TRAVEL LIFT Z)
G0 X125.1500 Y47.6500 (TRAVEL)
G0 X125.1500 Y47.6500 Z-15.2500 (TRAVEL LOWER Z)
M64 L0.7 (TURN TAMPER ON)
M11 (TURN OFF EXTRUDER SERVOING)
M3 S300 (TURN EXTRUDER ON)
G4 P0.0500 (DWELL)
M10 (TURN ON EXTRUDER SERVOING)
G1 F182.6948 S110.7241 X114.8500 Y47.6500 (INSET)
G1 X114.8500 Y37.3500 (INSET)
G1 X125.1500 Y37.3500 (INSET)
G1 X125.1500 Y47.6500 (INSET)
G1 F298.9551 S0.0000 X122.1500 Y47.6500 (INSET FORWARD TIP WIPE)
M65 (TURN TAMPER OFF)
M5 (TURN EXTRUDER OFF)
M66 L0.1991 (SET ACCELERATION)
G0 X122.1500 Y47.6500 Z-14.7500 (TRAVEL LIFT Z)
G0 X125.4900 Y47.9900 (TRAVEL)
G0 X125.4900 Y47.9900 Z-15.2500 (TRAVEL LOWER Z)
M64 L0.7 (TURN TAMPER ON)
M11 (TURN OFF EXTRUDER SERVOING)
M3 S300 (TURN EXTRUDER ON)
G4 P0.0500 (DWELL)
M10 (TURN ON EXTRUDER SERVOING)
G1 F182.6948 S110.7241 X114.5100 Y47.9900 (INSET)
G1 X114.5100 Y37.0100 (INSET)
G1 X125.4900 Y37.0100 (INSET)
G1 X125.4900 Y47.9900 (INSET)
G1 F298.9551 S0.0000 X122.4900 Y47.9900 (INSET FORWARD TIP WIPE)
M65 (TURN TAMPER OFF)
M5 (TURN EXTRUDER OFF)
M14 (INFILL SPINDLE ADJUSTMENT)
M66 L0.177 (SET ACCELERATION)
G0 X122.4900 Y47.9900 Z-14.7500 (TRAVEL LIFT Z)
G0 X124.8100 Y47.2600 (TRAVEL)
G0 X124.8100 Y47.2600 Z-15.2500 (TRAVEL LOWER Z)
```

```
M64 L0.7 (TURN TAMPER ON)
M11 (TURN OFF EXTRUDER SERVOING)
M3 S300 (TURN EXTRUDER ON)
G4 P0.0500 (DWELL)
M10 (TURN ON EXTRUDER SERVOING)
G1 F196.1646 S96.8836 X115.1900 Y47.2599 (INFILL)
G1 X115.1900 Y45.8999 (INFILL)
G1 X124.8100 Y45.9000 (INFILL)
G1 X124.8100 Y44.5400 (INFILL)
G1 X115.1900 Y44.5399 (INFILL)
G1 X115.1900 Y43.1799 (INFILL)
G1 X124.8100 Y43.1800 (INFILL)
G1 X124.8100 Y41.8200 (INFILL)
G1 X115.1900 Y41.8199 (INFILL)
G1 X115.1900 Y40.4599 (INFILL)
G1 X124.8100 Y40.4600 (INFILL)
G1 X124.8100 Y39.1000 (INFILL)
G1 X115.1900 Y39.0999 (INFILL)
G1 X115.1900 Y37.7399 (INFILL)
G1 X124.8100 Y37.7400 (INFILL)
G1 F298.9551 S0.0000 X127.8100 Y37.7400 (INFILL FORWARD TIP WIPE)
M65 (TURN TAMPER OFF)
M5 (TURN EXTRUDER OFF)

(BEGINNING LAYER: 3)
G0 W-0.4500 (MOVE W - LAYER CHANGE)
M1 (OPTIONAL STOP - LAYER CHANGE)

(Nozzle 0)
M12 (PERIMETER SPINDLE ADJUSTMENT)
M66 L0.2655 (SET ACCELERATION)
G0 X122.4900 Y47.9900 Z-14.7500 (TRAVEL LIFT Z)
G0 X125.8300 Y48.3300 (TRAVEL)
G0 X125.8300 Y48.3300 Z-15.2500 (TRAVEL LOWER Z)
M64 L0.7 (TURN TAMPER ON)
M11 (TURN OFF EXTRUDER SERVOING)
M3 S300 (TURN EXTRUDER ON)
G4 P0.0500 (DWELL)
M10 (TURN ON EXTRUDER SERVOING)
G1 F660 S400 X114.1700 Y48.3300 (PERIMETER)
G1 X114.1700 Y36.6700 (PERIMETER)
G1 X125.8300 Y36.6700 (PERIMETER)
G1 X125.8300 Y48.3300 (PERIMETER)
G1 F1080 S0 X122.8300 Y48.3300 (PERIMETER FORWARD TIP WIPE)
```

```
M65 (TURN TAMPER OFF)
M5 (TURN EXTRUDER OFF)
M13 (INSET SPINDLE ADJUSTMENT)
M66 L0.1991 (SET ACCELERATION)
G0 X122.8300 Y48.3300 Z-14.7500 (TRAVEL LIFT Z)
G0 X125.1500 Y47.6500 (TRAVEL)
G0 X125.1500 Y47.6500 Z-15.2500 (TRAVEL LOWER Z)
M64 L0.7 (TURN TAMPER ON)
M11 (TURN OFF EXTRUDER SERVOING)
M3 S300 (TURN EXTRUDER ON)
G4 P0.0500 (DWELL)
M10 (TURN ON EXTRUDER SERVOING)
G1 F660 S400 X114.8500 Y47.6500 (INSET)
G1 X114.8500 Y37.3500 (INSET)
G1 X125.1500 Y37.3500 (INSET)
G1 X125.1500 Y47.6500 (INSET)
G1 F1080 S0 X122.1500 Y47.6500 (INSET FORWARD TIP WIPE)
M65 (TURN TAMPER OFF)
M5 (TURN EXTRUDER OFF)
M66 L0.1991 (SET ACCELERATION)
G0 X122.1500 Y47.6500 Z-14.7500 (TRAVEL LIFT Z)
G0 X125.4900 Y47.9900 (TRAVEL)
G0 X125.4900 Y47.9900 Z-15.2500 (TRAVEL LOWER Z)
M64 L0.7 (TURN TAMPER ON)
M11 (TURN OFF EXTRUDER SERVOING)
M3 S300 (TURN EXTRUDER ON)
G4 P0.0500 (DWELL)
M10 (TURN ON EXTRUDER SERVOING)
G1 F660 S400 X114.5100 Y47.9900 (INSET)
G1 X114.5100 Y37.0100 (INSET)
G1 X125.4900 Y37.0100 (INSET)
G1 X125.4900 Y47.9900 (INSET)
G1 F1080 S0 X122.4900 Y47.9900 (INSET FORWARD TIP WIPE)
M65 (TURN TAMPER OFF)
M5 (TURN EXTRUDER OFF)

M5 (TURN EXTRUDER OFF END OF PRINT)
M65 (TURN TAMPER OFF)
M68 (PARK)
M11 (TURN OFF EXTRUDER SERVOING)
G1 W-16.75 (JOG W TO DOFFING LOCATION)

M30 (END OF G-CODE)
```

Index

Note: Page numbers followed by *f* indicate figures and *t* indicate tables.